Praise for
The New Carbon Architecture

Truly, what a fantastic, timely, important book!

— Paul Hawken, author of *Blessed Unrest* and *Drawdown*

I cannot overstate the importance of Bruce King's book at this
critical time. We know that in order to effectively address climate
change, we must go beyond building operations and address embodied
carbon — phasing out carbon emissions in building materials
and construction by mid-century; this book illustrates how.

— Edward Mazria, Founder / CEO Architecture 2030

That same carbon atom that's wreaking havoc in the atmosphere is
a building block for many great traditional and new building materials.
The New Carbon Architecture shows us how in ways that are both
practical and imaginative — truly a resource for our times.

— Nadav Malin, President, BuildingGreen, Inc.

Bruce King provides a valuable and unique reference for
understanding how one-fifth of all carbon emissions from buildings
are currently not being counted or even comprehended. Understanding
embodied energy and incorporating it into design thinking and
product development is the next frontier for green building practice.
I recommend this book as a primer for anyone interested in
combatting global climate change via building science.

— Jerry Yudelson, PE, LEED Fellow
"The Godfather of Green" — Wired Magazine
Author/Keynote Speaker/Sustainability Consultant

In *The New Carbon Architecture*, Bruce King delivers an emergent template for designing buildings in a future of climate uncertainty. The climate clock is ticking and we urgently need the ideas King and his colleagues present if we are to ensure comfort, safety, and resiliency in our next-gen built environment. The litany of "no regrets practices" King offers provides both adaptation and mitigation benefits in an industry not well known for offering either.

— David A. Schaller, retired EPA climate and sustainability coordinator

Bruce King and his crew of knowledgeable, enthusiastic authors have given us a great starting point for designing and (re-)creating our built environment. This is an important book for the entire design industry to read; from industrial designers and chemists to natural building craftspeople. It gives us all a starting point for the transformation of our infrastructure into one that is truly sustainable and healthy — while reducing the quantity of greenhouse gases in the atmosphere by simply using them as building blocks instead of emitting them. All this, and they show us that we can have fun doing it!"

— Tim Krahn, P. Eng., Structural Engineer, Building Alternatives Inc

THE NEW CARBON ARCHITECTURE

BUILDING TO COOL THE CLIMATE

BRUCE KING AND FRIENDS

new society
PUBLISHERS

Printed in Canada. Second printing, 2020.

Inquiries regarding requests to reprint all or part of *The New Carbon Architecture* should be addressed to New Society Publishers at the address below. To order directly from the publishers, please call toll-free (North America) 1-800-567-6772, or order online at www.newsociety.com

Any other inquiries can be directed by mail to:
New Society Publishers
P.O. Box 189, Gabriola Island, BC V0R 1X0, Canada
(250) 247-9737

LIBRARY AND ARCHIVES CANADA CATALOGUING IN PUBLICATION

King, Bruce (Structural engineer), author
    The new carbon architecture : building to cool the climate / Bruce King.

Includes bibliographical references and index.
Issued in print and electronic formats.
ISBN 978-0-86571-868-5 (softcover).--ISBN 978-1-55092-661-3 (PDF).--
ISBN 978-1-77142-256-7 (EPUB)

    1. Sustainable architecture. 2. Sustainable buildings--Design and
construction.
3. Dwellings--Environmental engineering. 4. Building materials--Environmental
aspects. I. Title.

NA2542.36.K56 2017   720'.47   C2017-906366-9

                                    C2017-906367-7

Funded by the Government of Canada / Financé par le gouvernement du Canada

New Society Publishers' mission is to publish books that contribute in fundamental ways to building an ecologically sustainable and just society, and to do so with the least possible impact on the environment, in a manner that models this vision.

THIS BOOK is a project of the non-profit Ecological Building Network, or EBNet, which has been developing and promoting healthy, low-carbon building for 20 years. ecobuildnetwork.org

... in collaboration with the Embodied Carbon Network, which was convened in 2016 to provide a mechanism for individuals and firms to connect, conduct research, and promote awareness of embodied carbon in the built environment.

carbonleadershipforum.org/embodied-carbon-network

for women
thank you

O NE CAN ASK what might it take to have an agriculture that does not degrade the soils, a fishery that does not deplete the ocean, a forestry that keeps watersheds and ecosystems intact, population policies that respect human sexuality and personality while holding numbers down, and energy policies that do not set off fierce little wars ... We know that science and the arts can be allies. We need far more women in politics. We need a religious view that embraces nature and does not fear science; business leaders who know and accept ecological and spiritual limits; political leaders who have spent time working in schools, factories or farms, and maybe a few who still write poems. We need intellectual and academic leaders who have studied both history and ecology and who like to dance and cook. But what we ultimately need most are human beings who love the world.

— Gary Snyder, *Back on the Fire* 2007

Contents

All text, photos and illustrations are by Bruce King unless noted otherwise.

Acknowledgments

First and foremost, profound thanks to the contributing authors to this book — for your many years of largely unrewarded and unappreciated work in this emerging field, and then for taking the time to provide your piece of this whole. The world is made better for your work. A particular shout-out to:

+ Ann Edminster for herding so many kitties, of both the human and conceptual kind, in defining and moving toward true net-zero architecture;

+ Ed Mazria for your tireless and effective work in alerting the world and our profession to the perils and promise of the built environment;

+ Kate Simonen for establishing, energizing, and guiding the Carbon Leadership Forum and its spinoff Embodied Carbon Network, and

+ Larry Strain for your untiring service and constant quiet leadership.

Many, many others have helped, directly and indirectly, in creating this book and in fostering and leading the emerging sciences of Life Cycle Analysis, Embodied Carbon, Biomimicry, just better building, and of course climate science itself. Thank you to:

David Arkin, Arup, Janine Benyus, Rachel Bevan, Marcial Blondet, Hana Mori Böttger, Nat and Sarah Cobb, Collins Products LLC, Columbia Forest Products, Denis Corbett, Sukita Reay Crimmel, Don Davies, Darrel DeBoer, Linda Delair, Kris Dick, David Easton, Everybody at Book Passage, Everybody at Peet's Coffee in Northgate, Everyday Zen Sangha, Terry Gamble-Boyer, Jacob Deva Racusin, Jittery John's Coffee, Leif Johnson, David Eisenberg, Pliny Fisk, Khosrow Ghavami, John Glassford, Min Hall (ILYM), Carol Hampf, Paul Hawken, Paul Holland, Interface, Bjørn Kierulf, Kenneth and Virginia King, Peter Kloepfer, Penny Livingstone-Stark, Ace McArleton, P. K. Mehta, Brad Roberts, Emily Rydell Niehaus, Graeme North, Elizabeth Patterson,

Kirsten Ritchie, Holmer Savastano Jr., Siegel & Strain Architects, Dawn Marie Smith, Nehemiah Stone, David and Nancy Thacher, John and Carry Thacher, Anni Tilt, Linda Toth, Carol Venolia, Cameron Waner, David Warner, John Warner, Margery and Steve Weller, Tom Woolley, Linda Yates, and all of you who belong on this list and are only omitted due to my lapse.

Final and special thanks to my wife, Sarah, who without being asked jumped right in and provided a thousand forms of support from start to finish. Without you this just wouldn't have happened. And to my children, Tyler and Lucy, I do it for you.

Any errors and omissions are mine.

Preface: Buildings Made of Sky

This would be easy if it weren't so hard.
— Yogi Berra

RECENTLY I WAS DRIVING IN AMERICA and pulled up at a stoplight behind a Tesla. You know Tesla, right? The latest and most talked about electric car, renowned for its power, handling, and just overall coolness. I drove one once, and can attest: it was great!

This particular Tesla had a license plate that read ZEROCARB — meaning, presumably, that the owner was proud of his zero carbon emissions car. No climate villain here! There's no way to know for sure, but I'd guess that this owner believed his claim, believed that his driving had no effect on the climate, unlike the rest of us bozos in our gas-powered stinkers.

As I sat there pondering for a few moments behind that license plate, this book was born. Because I was thinking what you're maybe thinking: "Huh?" That Tesla doesn't get recharged by twinkle fairies, and didn't appear by magic in the world — and neither do ovens or shoes or buildings. By some estimates, the energy required to make a smartphone, just for one example, is more than 70 times the energy it takes to charge it for a year — not to mention all the waste products, water, and emissions of many sorts that are involved. There is work and energy and rearrangement of some stuff into other stuff to make a Tesla — or a building. To believe otherwise is ignorance, to pretend otherwise is disingenuous and even somewhat dangerous. That Tesla moment sparked me to focus even more on the so-called *embodied carbon* of buildings, and on the many emerging technologies that will turn buildings from climate villains (which they now very much are) into climate champions that can soak carbon out of the air: buildings made of sky. The New Carbon Architecture was born, or at least conceived.

I might have been able to bang this thing together by myself, but where's the fun in that? For a quarter of a century, I have been noodling around with so-called *green* and *natural* and *alternative* building, and for that have been rewarded, if not with gold and silver, with friendships and correspondence with an extraordinary panoply of like-minded folks all over the world. If I had called on them all, this book would have hundreds of authors, so to stay sane, I kept it to the 20 or so who have really made a mark in advancing one or another aspect of the New Carbon Architecture. It saved me a lot of extra work and research, and results in a much better book for you, reader.

I guess I should add: this could have been a much bigger book. It might have been a dense 400-page tome, fully reporting the state of the art with tables, graphs, and other hallmarks of good science, or it could have been shaped as an academic textbook. But it seemed better to get the idea out into the world, as simply and readably as possible, beyond the few thousand people in the world already aware of this emerging and exciting notion of building with carbon. These colleagues of mine are not lightweights, and it took some persuading to get them to provide just the "elevator pitch" summaries of their respective work in their respective fields, laboratories, academia, and the unforgiving marketplace. If you want to dig deeper, we offer some pointers. Know that this is a very fast-developing subject that will look different — guaranteed — in a year or five or twenty. Consider this book to be a sort of heads-up, a shout-out for a very cool thing emerging on Earth.

Please enjoy, and please let us — the Ecological Building Network — know what you think: ecobuildnetwork.org

Introduction

*The primary task of any good teaching
is not to answer your questions,
but to question your answers.*

— Adyashanti, *The Way of Liberation*

IMAGINE:

YOU WALK INTO A BRAND NEW BUILD-
ING and immediately sense something
is different. The structure is all exposed
wood — columns, beams, even floor and
roof are all great curving slabs of timber elegantly joined together from smaller pieces.
The skin and insulation, which you can also
see, are straw bound into shapes that shed
rain and insulate walls. The foundation is soil
from the site transformed by invisible microbes into strong concrete to hold everything
up, and the warm, leatherlike floors need no
additional covering. It somehow looks like a
barn but smells like a forest, and feels more
like an inviting bedroom or an elegant mu-
seum. It's nicer than any building you've ever
been in before.

And it's not a handmade house in the
woods — it's a new downtown office building, nine stories high, full of people and filling
half a city block. It gathers all the power and
water it needs, is elegantly lit by daylight, and
processes all of its own water and wastes into

soil for the courtyard gardens. And, though
you can't see this, compared to what might
have been built a decade earlier, its construction put thousands of tons less carbon into
the air — and pulled hundreds more tons
out of the air to serve as its walls, floors, and
roof.

The New Carbon Architecture: a building made of sky. For the first time in history,
we can and should build pretty much anything out of carbon that we coaxed from the
air. We can structure any architectural style
with wood, we can insulate with straw and
mushrooms, we can make concrete — better concrete — with clay, microbes, smoke,
and a careful look in the rearview mirror and
the microscope. All of these emerging technologies and more arrive in tandem with the
growing understanding that the so-called
embodied carbon of building materials matters a great deal more than anyone thought
in the fight to halt and reverse climate
change. The built environment can switch
from being a problem to a solution. And it

1

doesn't matter whether or not you accept that climate change is anthropogenic: all the technologies described in the pages to follow make sense for a host of reasons, not least that they are much nicer buildings to occupy, and just happen to pull carbon out of the air.

But to back up a bit . . .

Human beings started building about eight thousand years ago with the dawn of the agricultural revolution, and that extended worldwide moment was arguably the most disruptive in history for us and the rest of life on Earth. Rather than hunt and forage about the landscape for our food, we grew it in one spot, and next thing you know, there was architecture, political states, wealth and poverty, Gutenberg and Einstein, global tension, Lady Gaga, and drive-thru WiFi-enabled hamburger stands in Cairo.

And billions more of us.

We've been developing the art and science of building for these thousands of years, mostly learning from trial and error, but as of the last few centuries also learning and developing via science. We know an awful lot more about how things work than we ever did, but can also dimly see how much we still don't know, such as what most of the universe is made of.

Speaking of what things are made of, in many ways the history of architecture follows the development of materials — the history of people messing around with things they found in the landscape to get bricks, then boards, then toilets, then building-integrated photovoltaic panels. People learned to fire clay to make pottery and bricks, and when the kilns were made of limestone, they discovered that the intense heat also changed the rocks: lime plaster, concrete, Pantheon. In some places the potters saw shiny metal come oozing out of certain heated rocks: copper, bronze, iron, Golden Gate Bridge. Two hundred years ago, the predecessors of modern structural engineers in England placed iron bars in newly invented Portland cement concrete, and architects went wild like they never could before: the Sydney Opera House and every downtown skyline in the world with lights, plumbing, and comfort hundreds of feet in the air. In some places people saw oil oozing out of the ground, then drying to tar: vinyl siding and the interstate highway system, not to mention plywood and air conditioning. And so on. Seems like the party would never stop, but of late the many large and hidden costs have come due, and we have to change not just the way we build, but what we build *with*.

Every modern industrial society has codified systems and materials of construction that are based on abundant fossil fuels, and on having an "away" where we can throw things. All the laws, standards, and codes are still rigidly based on doing things that way, even penalizing and inhibiting those who seek better ways to build. For the past century, it has been increasingly easy and cheap to extract, process, assemble, and transport everything we use in construction, but that won't last much longer. At this writing, in early 2017, fossil fuels are surprisingly cheap due to a variety of global conditions (*Peak oil? Are you kidding?*), so to warn of their limited supply seems ludicrous. But the climate is definitely changing, and the effects are arriving harder and faster than we expected even ten years ago. The "heat, beat & treat" approach to making and processing materials is killing us, as is the notion that we can throw anything we want into landfills, water, soil, or air, because building materials account for about 10 percent of global carbon emissions and 25–40 percent of solid wastes. That just has to change. We have a new ball game.

Some of us who design and build have lately started noticing that Nature builds all sorts of things, and has been doing so for the four billion years of life on Earth. She has a hell of a head start on the trial-and-error path; maybe we can and should peek over her shoulder and see if we can't cheat a bit. How does a mussel build its shell? How do spiders spin their webs? How does a redwood tree stand and remain very much alive at 380 feet — and why doesn't it grow higher? How do birds stay warm and dry at night?

When facing design challenges from the small (How can I illuminate a surface or keep out rain?) to the large (Can nine billion human beings live on Earth without wrecking everything for themselves and the other critters, maybe even be a welcome presence?), we might ask: *What would Nature do?*

Some simple and semi-obvious things come right to mind: Nature runs on solar and geothermal energy with no other external energy inputs, and Nature uses what is at hand either by growing it like a clam grows its shell, or harvesting nearby resources as birds do for their nests. There's no FedEx, there's no power grid, there are no artificial chemicals to worry about.

But you and I live in a highly interdependent industrial society, where the sudden disappearance of FedEx, the power grid, a huge multitude of problematic chemicals, and all the other trappings large and small of modern life, would make for a whole lot of suffering for a whole lot of people. We've built a better life for more and more of us, but at the same time made quite a mess, so can we clean it up? Can we wean ourselves off of the fossil fuel habit? This ship doesn't turn very fast, but can we plot a course to a world that works for everybody?

Sure. Technologically, we're scarily clever creatures. It took less than two and a half years between Franklin Roosevelt authorizing the Manhattan Project and the first atomic explosion in the New Mexico desert (for better or worse). It took only eight years between John Kennedy's proclamation and Neil Armstrong's foot stepping onto the Moon's surface. And both of those projects were designed and executed by men and women using slide rules, unreliable wire telephony, and computers far less powerful than the average laptop of today. When we collectively set ourselves to do something, for better or worse, we tend to get it done. Of late there's been plenty of the better but also far too much of the worse. How about let's change that, and get more better and less worse.

This book offers a few suggestions for a more-better built environment, not so much a road map as a collection of useful essays sketching a new palette of materials for a new century. "Net-zero" buildings that use less energy than they generate are a good start, but don't go nearly far enough; here we point out how to design and build truly zero carbon buildings: the New Carbon Architecture.

10%

How much impact might this make on climate change? That would be a rich and nuanced topic for a graduate level thesis, and we hope someone takes up the challenge. But the short answer is: a lot. According to the United Nations Environment Programme, "Though figures vary from building to building, studies suggest that . . . generally 10 to 20 percent of [global] energy is consumed in materials manufacturing and transport, construction, maintenance and demolition."[1]

Various and multiple other studies assign building materials 5 to 15 percent of global emissions, there being no consistent methodology nor data sets to draw from. Call it 10 percent of global emissions, and there's your impact. We propose to reduce that number to zero — and then beyond by a new "carbon positive" architecture that builds with the carbon enticed from sky. We are in technological reach, within a generation, of a global construction industry that is not only "Net-zero," generating more energy than it needs to operate, but in its materials pulls more carbon out of the air than it puts up. We can reverse the emissions engine.

I suppose it bears noting that we the authors are unabashed materials geeks (among other talents), but we're not dense. We recognize that the materials of architecture are not the only component of climate-friendly design, much less of climate work writ large. But we do want to make clear that carbon sequestering architecture is an essential component among the many emerging technologies and strategies for climate *cooling*, from energy to transportation to waste management to water. In particular, we have a keen eye on agronomy and the study of soils, and all the gazillions of amazing little creatures therein, for it's starting to look like that's where we will find real wealth and the wisdom to grow food, clothing, and shelter in fantastic, lovely, and healthy new ways — not to mention sequester stupendous amounts of carbon. We take pride and delight in joining the broader climate effort, and hope you will find useful the news we bring and the vision we share. It's a whole new and lovely ball game.

A Word about "Carbon"

I know you believe you understand what you think I said, but I'm not sure you realize that what you heard is not what I meant.

— Richard Nixon

Carbon. It's a good thing. Right up there, Number 6 in the periodic table, and one of the most common elements on Earth. Carbon is here because a very, very long time ago uncounted millions of first-generation stars created it by nuclear fusion in their cores, then offered it by supernova explosion to the universe. Along with all sorts of other elemental fusion dust, it floated around, eventually to condense by gravity into planets and the world we know. And, as many have noted, it is the party animal of elements: it loves to bond with things like nitrogen, iron, hydrogen, and oxygen to make all sorts of interesting delights such as giraffes, redwood trees, poodles, and you. You read these words with carbon eyes, and hold this book with carbon hands. Please enjoy; not every blob of stardust gets to be conscious for a brief few moments under the sun and run around on a lovely planet with all sorts of other delightful carbon blobs. Congratulations, you lucky dog!

Carbon is a good thing, but too much of anything in the wrong place becomes pollution, or even poison. This book is but one of thousands of efforts to reverse the increase of gaseous carbon in the air, which is disrupting the climate in ways that we can't fully predict, and so far mostly don't like. So we enthusiastically join the growing conversation for climate solutions, but must first be clear about the terms we use. *Carbon* is bandied around a lot, but people often mean slightly different things by it.

3.67

Carbon and *carbon dioxide* (CO_2), for example, are two different things, though they get interchanged quite a lot in climate conversations. The fraction of carbon in carbon dioxide is the ratio of weights: the atomic weight of carbon is 12 atomic mass units, while the weight of carbon dioxide is 44 because it includes two oxygen atoms that each weigh 16. You switch from one to the other with this formula: one ton of carbon is equivalent to $^{44}/_{12} = 3.67$ tons of carbon dioxide. (Methane, or CH_4, a major greenhouse gas with 86 times the warming

potential of CO_2, has an atomic weight of 16, so the ratio is less pronounced: a ton of carbon in your building equals $^{16}/_{12} = 1.33$ tons of methane in the air.) Plants like straw (about 35–50 percent carbon) or softwoods (about 50 percent carbon) *sequester* (that is, durably store) carbon by absorbing carbon dioxide and releasing the oxygen. They feed us oxygen with their respiration, and we oxygen-breathing creatures feed them CO_2 with our respiration. Cool deal, huh? A ton of carbon in the forest or field — or as part of a building — represents or simply is 3.67 tons of carbon dioxide absorbed from the air.

Also, following convention, we will sometimes use *CO_2e* to denote carbon-*equivalent* emissions from carbon and other gases such as methane, calibrated according to each one's *global warming potential* (GWP) because some gases have ten or a hundred or even thousands of times the heat-trapping effect of carbon dioxide. Chapters One and Two will define and expand on what we mean by *embodied carbon* aka *carbon footprint*, but from here on out, we'll use those terms to connote embodied carbon equivalents, or eCO_2e. We might also sometimes be lazy and just say "carbon" when we mean CO_2e emissions, but we trust you'll get the drift without confusion.

Finally: *embodied energy* and *embodied carbon*. Be warned that terms like *zero energy* (aka ZE), *net-zero energy* (aka NZE), *zero net energy* (aka ZNE) are all increasingly tossed about in loosely interchangeable ways in conversation around building energy efficiency. Even more confusing, their close cousins *zero carbon* and *zero net carbon* are also appearing more frequently. This is a rather complex matter in itself, as terms change meaning with scale (product, building, community, nation, or globe?), with grid efficiency (coal, hydro, nuclear, wind? etc.), time frame (daily, annualized, or lifetime?), and other factors. In the pages that follow, some authors will variously use embodied energy and embodied carbon, and for our purposes those are in tandem; that is, though the units for measurement are different, they rise or fall roughly in parallel. (In Chapter Two: Counting Carbon, we discuss how they can diverge, as when products are manufactured with electricity from a coal-dependent grid vs. a hydropowered grid.) The growing consensus is that zero carbon (vs. zero energy) should be our societal goal across all industry, and so we will favor that term from here on out. Even better, we will also sketch out the possibility of a *carbon positive* architecture defined by more carbon sequestered than is ever emitted.

A book made of carbon, written by carbon, for carbon, on how to build carbon shelter to protect us from a sometimes hostile carbon planet.

Shall we dance?

Notes

1. *Buildings and Climate Change*, United Nations Environment Programme, 2009.

Chapter One:

Beyond Zero: The Time Value of Carbon

by Erin McDade

A Global Carbon Limit

In December of 2015, the world came together in Paris for the United Nations' 21st Conference of the Parties (COP21), and signed the historic Paris Climate Agreement. This agreement commits almost 200 countries to helping limit global temperature increase to "well below 2°C above pre-industrial levels and to pursue efforts to limit the temperature increase to 1.5°C."

These temperature increase limits are in response to the international scientific community's widely accepted two-degree Celsius tipping point. Global temperatures have been increasing steadily since the industrial revolution, but the scientific community believes that if we can peak our global increase and begin to cool the planet before we gain two degrees, the effects of climate change will be reversible. In other words, if we meet this target we can return the planet to pre-industrial conditions. However, scientists believe that if we pass that two-degree threshold, the effects of climate change will begin to cascade, spin out of control, and become irreversible.

Buildings Are the Problem; Buildings Are the Solution

In addition to the historic signing, COP21 made history by hosting its first ever Buildings Day in recognition of the crucial role that the building sector must play in reducing global CO_2e emissions. The US Energy Information Administration (EIA) estimates that constructing and operating buildings accounts for nearly half of all US energy consumption and fossil fuel emissions. Globally, cities consume nearly 75 percent of the world's energy, mostly to build and operate buildings, and cities are responsible for a similar percentage of global emissions. The building sector is a significant part of the climate change problem, but this also means that if we can eliminate carbon emissions from the built environment, we can significantly reduce overall emissions, ameliorating and potentially even solving the climate change crisis.

Zero by 2050

According to the United Nation's Inter-governmental Panel on Climate Change (IPCC), the organization that hosts the annual Conference of the Parties, global temperatures have already increased by 0.85 degrees Celsius since pre-industrial times, meaning that we're nearly halfway to our maximum temperature increase threshold. In order to predict future temperature increases as they relate to fossil fuel emissions, the IPCC periodically publishes projections based on a number of global patterns, ranging from business as usual to aggressive emissions reductions. In 2013, prior to the Paris Climate Agreement, the IPCC published four projection scenarios. The business-as-usual scenario, in which our consumption of fossil fuels continues to grow exponentially, projected that we would pass the two-degree tipping point around 2040. Even the more aggressive reduction scenarios, in which global emissions peak between 2050 and 2080 and then begin to diminish, showed us passing a two-degree increase near 2050. While an extra decade below two degrees would certainly be an improvement, following these projections would simply be delaying the inevitable — climate change would still spin out of control and become irreversible. The fourth and most aggressive scenario published, in which global emissions peak and begin diminishing in 2020, gave us our best chance of staying below the two-degree Celsius threshold, but unfortunately still predicted a large chance of surpassing that tipping point.

In response to the Paris Climate Agreement's aggressive target of a 1.5-degree

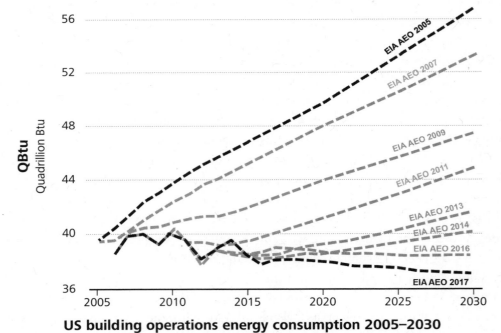

US building operations energy consumption 2005–2030

Data source: US Energy information Administration, Annual Energy Outlook (EIA AEO)
© ARCHITECTURE 2030, 2017

Fig 1.1.

maximum increase, the IPCC published an additional emissions scenario that gives us an 85 percent chance of staying below a two-degree increase. However, this scenario requires global carbon emissions to peak immediately, and for us to fully phase out our use of fossil fuels, in every sector, by mid-century. This means that in order to meet the targets set forth by the Paris Climate Agreement, the global building sector must be carbon free by the year 2050.

The Zero Net Carbon Gold Standard

Since the beginning of the green building movement in the 1970s, the design community has focused mainly on increasing the efficiency of *operating* our buildings —

reducing the energy consumed (and carbon emitted) in keeping everyone warm (or cool), keeping the lights on, etc. Both technology and design have improved drastically in the last 50 years, and now Zero Net Carbon (ZNC) is the gold standard for sustainable construction. A ZNC building is a highly efficient structure that produces renewable energy onsite (typically using photovoltaics), or procures as much carbon-free energy as it needs to operate. ZNC buildings are being constructed globally in almost all climate zones, space types, and sizes, proving the viability of this standard, and their reduced carbon emissions are being documented.

Each year the US Energy Information Administration publishes the Annual Energy

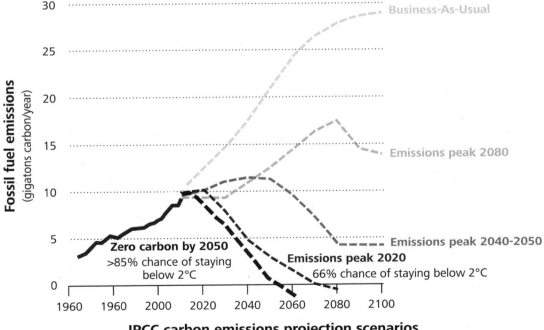

IPCC carbon emissions projection scenarios

Source: IPCC 2013, Representative Concentration Pathways (RCP); Stockholm Environment Institute (SEI), 2013, Climate Analytics and ECOFYS, 2014. Note: Emissions peaks are for fossil fuel CO_2-only emissions.
© ARCHITECTURE 2030, 2017

Fig 1.2.

Outlook, predicting future US energy demand based on current consumption trends. In 2005, before Architecture 2030 launched the 2030 Challenge and catalyzed the US and global building communities to begin targeting zero operational emissions by 2030, the EIA's projections predicted exponential growth in building operational energy consumption. Following the launch of the 2030 Challenge, each subsequent year's projections showed a decrease in energy consumption, and projections flatlined in 2016. The EIA's 2017 projections predict that US building operations will consume less energy in 2030 than in 2005, despite consistent and significant growth in the building sector. With this downward trend, and the increasingly frequent construction of ZNC buildings, the world seems on track for meeting the widely adopted commitment to zero operational carbon emissions by the year 2030.

Embodied Carbon: Getting to Real Zero

However, the emissions resulting from operating our buildings only represent one side of the coin. In fact, even before a building is occupied and any energy has been used for operation, the building has already contributed to climate change — usually in a significant way. These mostly unnoticed effects are the result of the construction process itself, and include emissions resulting from manufacturing building products and materials, transporting them to project sites, and construction. We refer to this as *embodied energy* (energy consumed pre-building operation) or *embodied carbon* (carbon emitted pre-building operation). To date, even within the green building community, these emissions are usually ignored in the conversation about the building sector and climate disruption.

On day one of a building's life, one hundred percent of its energy/carbon profile is made up of embodied energy/carbon. Embodied carbon emissions end upon the completion of construction, while operational carbon is emitted every day for a building's entire life. (Well, not quite. All buildings get maintained, painted, reroofed, remodeled, added to, repaired, and so on, causing embodied carbon to continue to climb slightly in short bursts. But for most buildings this effect is minor by comparison, so for this discussion we treat embodied carbon as just that from the original construction.) Over the life span of a typical building, the cumulative operational emissions almost always eclipse the initial embodied ones, and by the end of the building's life, embodied energy accounts for only a fifth or less of the total consumed by the building. Even if that same building is constructed to operate twice as efficiently, cumulative operational emissions are still greater than initial embodied ones. (See Figure 1.3.)

Since embodied energy accounts for an average of 20 percent of a building's total energy consumption over its life, it is understandable that the building sector's historic focus has been on operational (instead of embodied) energy and carbon. However, with the signing of the Paris Climate Agreement committing the world to a carbon-free built environment by 2050, we clearly can no longer ignore embodied carbon. In fact, new research indicates that to date we have significantly underestimated the significance and time sensitivity of embodied carbon in overall building sector emissions.

Emissions Now Hurt More than Emissions Later: The Relative Importance of Embodied Carbon

Over a typical building's 80–100-year life span, operational emissions dwarf embodied emissions. But if our deadline for eliminating building sector emissions is three decades or less, the timeline is much shorter and the relative importance of embodied carbon changes. Assuming a building is constructed today and operates 50 percent more efficiently than a typical building, by 2050 only 45 percent of the energy consumed by that building will have been used for operations, meaning that 55 percent of that building's total energy consumption is embodied energy. And the closer to 2050 the building is constructed, the more embodied carbon emissions eclipse operational carbon emissions. Furthermore, as we target ZNC for all new construction and buildings are designed to meet increasingly rigorous performance standards, the amount of operational carbon emitted decreases and is eliminated, meaning all of a building's carbon emissions are the result of embodied carbon. (See Figures 1.4 and 1.5.)

Embodied Carbon in the Future

Between 2015 and 2050, more than two trillion square feet of new construction and major renovations will take place worldwide,[1] the equivalent of building an entire New York City (all five boroughs) every 35 days, for 35 years straight! If the built environment is to be carbon free by 2050 and meet Paris Climate Agreement targets, how we in the design community design and construct this two trillion square feet, and how we value and evaluate its embodied carbon, is crucial.

Even conservatively assuming that all of this new construction operates twice as efficiently as typical construction, between now and 2050, 80–90 percent of its energy profile will be made up of embodied, not operational, energy. The carbon math is similar though not identical due to variations in grid energy emissions.

This isn't to say that operational performance isn't important. Barring major renovations, a building's operational emissions patterns are locked in on day one: an inefficient building constructed today will probably still be inefficient in 2050. And while major renovations, in which we upgrade to Zero Net Carbon standards, are an important part of decarbonizing the built environment, each renovation requires new building materials and more construction, which further increase emissions. (See also Chapter Three: Rebuild, by Larry Strain.) It is crucial that we consider immediate embodied carbon impacts when constructing this massive additional building stock. Sometimes, that may even mean valuing lower embodied carbon strategies, or better, using carbon-sequestering materials as presented in the rest of this book, over carbon-intensive strategies that only minimally improve operational performance.

The Time Value of Carbon

Without a deadline, we might continue to dismiss embodied carbon impacts as minimal compared to operational carbon. But climate change is urgent: all carbon emissions must

Who cares about embodied carbon?
Operational emissions are far bigger, right?

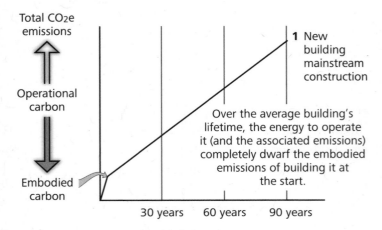

Fig 1.3: *For the first decades of green building, no one thought that embodied carbon mattered very much.*

Moving from mainstream *to* efficient *to* net-zero
BIG improvements!!!

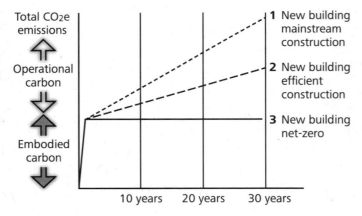

Fig 1.4: *Most of "green design" to date is about reducing operating energy (or carbon emissions).*

be eliminated from the built environment by the year 2050. In traditional analyses of embodied vs. operational carbon, which consider the building's whole life span, rarely are initial embodied carbon outputs as impactful as the operational savings that they allow — high-performance design savings almost always pencil out as worth the embodied carbon investment. But as high-performance construction becomes the standard, and designers increasingly work to eke out every last drop of operational savings, that little bit of improved performance often won't balance out the required embodied carbon investment for decades — well past our 2050 deadline. Therefore, any carbon reduction strategy we consider must be evaluated based not only on potential savings, but also on how quickly those savings can be achieved. If, for example, adding fixed shading to a new building will improve its performance so much that the increased operational savings exceed the added materials emissions in at most a few years, that may be a smart strategy. But if adding an extra inch or two of carbon-intensive, high-density insulation improves efficiency by a few percentage points, but it takes 50+ years for the operational improvements to outweigh the embodied costs, is that the right choice? (Hint: No.)

To have any hope of meeting our climate change goals, we must rethink our traditional carbon analysis mechanisms and design processes. Whole building life spans do not accommodate the urgency of climate change; carbon emitted today has much, much more impact than carbon emitted after 2050, and we can't continue to underestimate the effects of embodied carbon emissions. If our remarkable success in high performance design continues, embodied carbon may well prove to be our downfall — or the key to solving climate change. It's up to us to decide.

The effect on the climate is = time x emissions
The *impact* is the grey area under the curve

Emissions are hugely amplified by *when* they occur — embodied carbon is greatly weighted, very much like the time value of money.

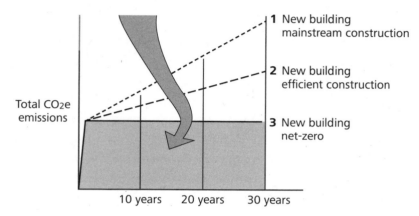

1 New building mainstream construction

2 New building efficient construction

3 New building net-zero

Total CO2e emissions

10 years   20 years   30 years

Fig 1.5: *But, oops! We're suddenly realizing that embodied carbon matters a lot, and is about half to three-fourths of the climate impact of your next project over the next two decades.*

1, 2, and 3 all have *big climate impacts* because they emit carbon right from the start

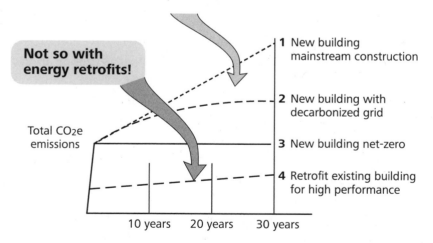

Not so with energy retrofits!

1 New building mainstream construction

2 New building with decarbonized grid

3 New building net-zero

4 Retrofit existing building for high performance

Total CO2e emissions

10 years   20 years   30 years

Fig 1.6: *The good news is, we can have both: low to zero operating and embodied emissions.*

Zero Energy in a Nutshell

by Ann Edminster

The US Department of Energy defines a zero energy building as "an energy-efficient building where, on a source energy basis, the actual annual delivered energy is less than or equal to the on-site renewable exported energy."[2] Notably, embodied energy is not included in this widely used definition; nor is transportation energy, nor are other components of personal energy use. Nonetheless, leaders in the green building community have been concerned with these other components for a long time.[3]

Zero energy (aka ZE, net-zero energy, NZE, zero net energy, ZNE) building as a concept has been in play in the energy efficiency community for approaching two decades, with attention increasing following Governor Arnold Schwarzenegger's 2005 executive order decreeing that by 2020, all new California homes would be zero energy, and by 2030, all the State's new non-residential buildings would follow suit. As 2020 has gotten closer, the rate of activity in the ZE arena has been accelerating. Today, ZE built projects and policy initiatives exist all around the globe, with California still at the forefront of ZE in North America, closely followed by the US Northeast.[4]

Passive House (Passivhaus), another burgeoning energy efficiency movement, is rapidly converging with ZE, as building professionals increasingly adopt the Passive House framework to facilitate attaining the high level of efficiency needed to achieve ZE.

Zero energy leaders have long acknowledged that ZE at the community scale is their real target, with focus shifting more recently to zero *carbon* communities. This has a number of far-reaching and challenging technical and policy implications related to electrification (elimination of natural gas, fuel oil, and propane in building operations), transformation of the utility grid, and aggregating energy demand and renewable energy production of individual buildings in order to achieve zero carbon operations at community scale.

At the same time, there is still much to be learned about achieving zero energy at the building scale, including factoring in embodied energy/carbon. By and large, the energy efficiency community still believes that embodied energy is of less concern than operating energy, because it is a smaller portion of the energy pie. However, the authors of this book, among others, have come to realize that with the element of time of utmost importance in the climate change equation, more attention to embodied energy/carbon from the ZE community is overdue. Some groundbreaking projects are beginning to tackle this challenge, notable among them the redesign of Terminal 1 at San Francisco International Airport, where the design team is analyzing embodied carbon of various design alternatives alongside operating energy, to enable them to make informed decisions about the overall energy/carbon performance of the design options.

Notes

1. IEA. 2016. Energy Technology Perspectives 2016, IEA/OECD, Paris.

2. DOE has formulated companion definitions for ZE campuses, portfolios, and communities.

3. As one such example, Dr. Raymond Cole, University of British Columbia, in the mid-2000s proposed a schema for personal energy uses that comprised household operating energy, routine personal transportation energy, the energy embodied in durable goods, in the food we eat, and in our vacation activities.

4. Net-zero Energy Coalition. 2016. *To Zero and Beyond: Zero Energy Residential Buildings Study.*

Chapter Two:

Counting Carbon: What We Know and How We Know It

by Catherine De Wolf, Barbara Rodriguez-Droguett, and Kathrina Simonen

Building Carbon Neutral

DO CARBON NEUTRAL BUILDINGS exist? In many instances, people will claim a building is "carbon neutral" if the building is net-energy neutral, using methods such as solar panels to generate power and passive solar to heat and cool. However, as this book highlights, building materials and products contribute significant carbon impacts over a building's life span. We argue that for a building to be truly carbon neutral, it would need to demonstrate total carbon neutrality through both net-zero embodied and operational carbon impacts, a difficult, but not impossible, challenge to achieve.

Embodied carbon, the greenhouse gases emitted by extracting, producing, transporting, using, and waste-treating materials, is commonly known as the *carbon footprint*. The embodied impacts of buildings are directly related to materials: both the *types* of materials chosen (as outlined in other chapters) and the *quantities* of materials used. (See also Chapter Ten: Size Matters.)

Evaluating the total embodied carbon of a building is typically done using *Life Cycle Assessment* (LCA). LCA is a calculation method that integrates data about the amount and types of materials and energy used, the manufacturing processes and associated chemical reactions, with data about emissions for each of these processes. An LCA reports the known environmental impacts resulting from these emissions. LCA data for different products and materials is typically developed by individual manufacturers or trade organizations and is not always publicly accessible. Aside from embodied carbon, LCA identifies a number of other environmental impacts, such as depletion of the ozone layer, creation of smog, and pollution of water, as well as toxicity to human health.

Whole Building Life Cycle Assessment (WBLCA) combines an estimate of the quantity and types of materials and products used to construct a building with embodied carbon (and other impact) coefficients to estimate the total embodied impacts. A rapidly growing number of tools to conduct both material and whole building LCAs are now available. This chapter draws upon the results of LCA studies to summarize what we know about whole building embodied carbon, and to highlight best opportunities for innovation and reductions in building embodied emissions.

At present, calculating the embodied carbon coefficients of materials isn't easy. Data quality is often outdated, incomplete, inconsistent, and/or germane only to its region of origin. For example, the newer kilns in China make "cleaner" (lower emissions) cement than the older kilns in the United States.[1] Lumber imported from Scandinavia has much higher transport emissions when the building is located in Africa compared to Northern Europe. Also, building materials can be produced in many ways. For example, steel can be produced from iron ore to make primary steel or from scrap steel in secondary steel production; the primary steel requires much more energy than the secondary steel. When comparing construction materials, the scope, life cycle stages and environmental impacts can be different from one study to another, and there is a high variability because of regional differences, improved efficiencies of production, new ingredient composition, and/or various mechanical properties required.

Within the growing LCA community, life cycle stages are defined as follows. Embodied carbon of materials measuring only material extraction and manufacturing is called Cradle-to-Gate as the analysis stops when the product leaves the "gate" of the manufacturing facility. When it also includes delivery to the construction site, it is called Cradle-to-Site. Adding the eventual repairs and replacements as well as the possible demolition and removal scenarios results in Cradle-to-Grave values. Including the benefits or loads that can occur through landfilling, reuse, recycling, or recovery assesses the Cradle-to-Cradle impact.

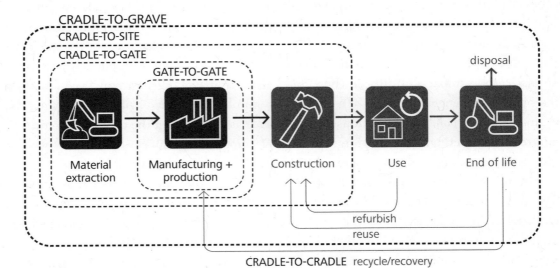

Fig. 2.1: *Adapted from Simonen (2014)*[2]

Considering the many variables of different regions, boundary conditions, and life cycle stages, we need to define a clear standard for comparing products to one another. Several countries have responded by starting national Environmental Product Declaration (EPD) databases based on voluntary contributions from their different material manufacturers. An EPD is similar to the mileage sticker on a car — it reports the expected environmental performance of a material or product. If your concrete, steel, or lumber producer has performed an LCA of their production process or the full life cycle of their products, they can write an EPD. The aim here is to be transparent and communicate information that can be compared to that of other products. The International Organization of Standardization (ISO) wrote rules to make sure these EPDs follow the same framework, and the industry is slowly building a global database with regional EPDs; this will allow designers to choose materials that have the lowest impacts and are most appropriate to their building project and location.

The Relative Impact of Embodied Carbon in Typical Buildings

As Erin McDade argues in Chapter One, carbon emissions from building construction are significant, and for individual buildings the relative importance of reducing embodied impacts is far more crucial than the green building community has recognized so far. In order to meet the Paris Climate Agreement,

Editor's Note

The growth of industry-supplied EPDs is an essential step toward a new carbon architecture — and it so far exhibits the same blindness of modern building codes and construction writ large. That is, only industry-supplied data is developed and recognized, omitting products or materials without the financial backing to "show up," both figuratively and literally. There is a large family of historic and indigenous building systems now enjoying a renaissance under the moniker "natural building," such as straw, clay, and bamboo (as discussed later in this book), as well as recycled products like tires, bottles, and shipping containers that are also appearing in ever-new and ingenious ways as human shelter. Historically, these materials have been largely confined to the rural and the poor; in modern art and literature, "mud huts" and "grass shacks" are regularly used to signify poverty. But those materials are no less effective for being free or cheap where they are nearby and plentiful. That is, they are ultra-low carbon, and will hopefully receive more careful consideration by designers and builders as the new carbon architecture evolves. As the following pages will describe, ultra-low carbon materials can do much more than people think, and are by no means constrained to low-end, low-rise construction. As software providers work toward the goal of having a constantly updating carbon tally running on a designer's computer screen, let it truly show *all* the options — not just the ones someone paid for. The same plea goes out, of course, to the academic, codes and standards worlds: recognize and work with the full palette of materials that people actually use.

the next 30 years are the critical time frame for action. Typical building assessments use a life span of 50 to 80 years to evaluate life cycle performance, but we argue that we should focus on the near term where reductions to embodied carbon are usually much more important than operating emissions.

To estimate the embodied carbon of a building, we must first define what we mean by "the whole building." In most LCA studies to date, this includes the *structure* (foundation and framing), and exterior walls and roof (or *enclosure*) and often includes interior walls and finishes. Only rarely are additional components such as plumbing and lighting fixtures, mechanical and electrical components, or exterior paving and other site work included. Figure 2.2 represents a relative contribution of initial embodied impacts (Cradle-to-Site) to construct a high-rise residential tower.[3] For this building, the structural system is about a third of the total impacts followed by the envelope (glass and aluminum are both energy-intensive

materials) and the interior finishes. In this study, we found that about 25 percent of the total initial impacts can be attributed to components that are often not included in LCA studies. Other studies of larger buildings show similar distribution of impacts. For projects on large suburban sites or with complex systems, the rarely tracked site work and system embodied impacts may be significantly higher than these estimates.

The above chart is only for the impacts up to the first day of construction. But what about the operating impacts? Studies have shown that over the life span of a building the impacts embodied in materials and products account for between 20 and 100 percent of the total lifetime emissions. Results vary depending on building use, location, material palette, and assumptions about the service life and future energy supply.[4] Why is the range so large? For zero net energy buildings, there are no impacts from operating the building, so all of the impacts result from materials. For conventional buildings over an 80-year life span, the operating impacts dominate, typically around 80 percent of the total. Figure 2.3 is a generic plot of the total carbon emissions of several different building options over a 30-year time frame. Embodied carbon is represented by the vertical increase at the far left representing the initial construction (renovations and maintenance are not included here for simplicity). The sloping lines on the chart represent operating impacts, which increase over time. The less efficient the building, the steeper the line because the rate of emission is higher. The solid lines assume the energy grid remains

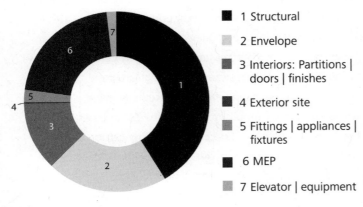

1 Structural

2 Envelope

3 Interiors: Partitions | doors | finishes

4 Exterior site

5 Fittings | appliances | fixtures

6 MEP

7 Elevator | equipment

Fig. 2.2: *Relative contribution of initial embodied impacts (cradle-to-site) to construct a high-rise tower.* Credit: Adapted from K. Simonen, *Testing Whole Building LCA: Research and Practice*

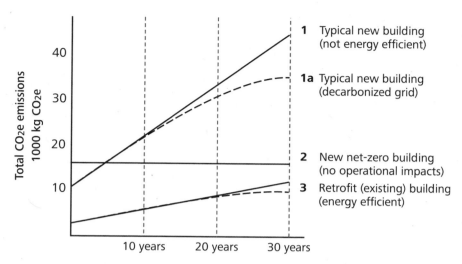

Fig. 2.3: *Total carbon emissions of several building options over a 30-year time frame.*
Credit: Adapted from K. Simonen, *Testing Whole Building LCA*

unchanged, while dashed lines assume we reach carbon neutral grid power within 30 years and thus the slope of the operational energy impacts flattens out over time.

These studies strongly support the argument presented by Larry Strain in Chapter Three: the lowest impact building for this crucial next 30 years is an existing building, retrofit to be energy efficient. There is typically much more bang for the buck in deep energy retrofits than in brand-new net-zero buildings.

Over the life span of a building, the climate impact of materials and products occurs mostly at initial construction, then in intermittent and relatively minor amounts during refurbishment, maintenance and repair, as well as at end of life. When embodied impacts are analyzed over a long life span of a building, some materials, most notably glass and interior finishes, become increasingly significant. While glass does not typically wear out, current assumption is that windows will be replaced to increase energy efficiency or because of broken seals in double-paned windows. When replaced, perfectly good glass is usually broken up at the jobsite and recycled into lower-grade glass products. We are designing buildings and their components with planned obsolescence, a habit that is becoming decreasingly viable. Designers looking to reduce the overall life cycle impacts should recognize that buildings change constantly and consider designing for durability and material/component reuse.

Comparing Structural Materials

Structure comprises the greatest weight in buildings and typically accounts for 30–50 percent of their embodied carbon emissions;[5] structural engineers have an important role to play in reducing the global warming potential (GWP) of the building sector. Given that the structural system is the primary contributor to a building's total embodied carbon, many studies have been conducted to determine if one structural system is environmentally preferable to another. In industrialized areas, the three main structural materials are concrete, steel, and wood, so

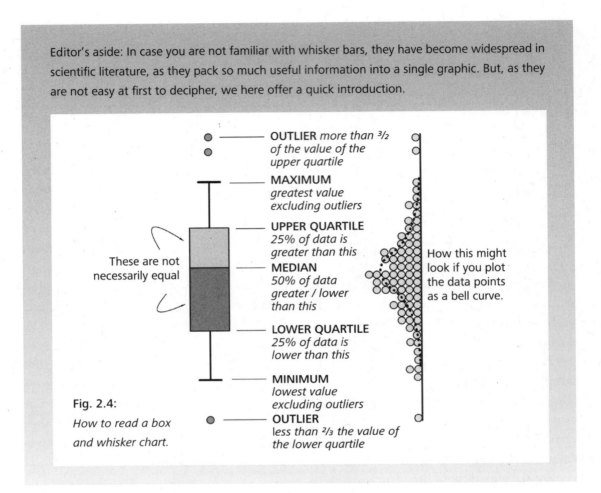

Editor's aside: In case you are not familiar with whisker bars, they have become widespread in scientific literature, as they pack so much useful information into a single graphic. But, as they are not easy at first to decipher, we here offer a quick introduction.

OUTLIER *more than* 3/2 *of the value of the upper quartile*

MAXIMUM
greatest value excluding outliers

UPPER QUARTILE
25% of data is greater than this

These are not necessarily equal

MEDIAN
50% of data greater / lower than this

LOWER QUARTILE
25% of data is lower than this

MINIMUM
lowest value excluding outliers

OUTLIER
less than 2/3 *the value of the lower quartile*

How this might look if you plot the data points as a bell curve.

Fig. 2.4:

How to read a box and whisker chart.

people naturally ask which is better — concrete, steel, or wood? The answer, as always in construction, is "it depends." At first glance, comparing just the *Embodied Carbon Coefficients* (ECC), expressed as emissions per unit weight, steel looks the worst: concrete ranges around 0.15 kgCO$_2$e/kg, steel 1.5 kgCO$_2$e/kg, and lumber only 0.1 kgCO$_2$e/kg. But this is quite misleading because of enormous variability in architectural designs, and the range of utility offered by a kilogram of different materials. One kilogram of steel achieves different things than one kilogram

of concrete or lumber. If we looked at emissions per volume, cost, or load-bearing capacity instead of per weight, the numbers would look very different. Also, the values for the embodied carbon of the different materials depend on the region in which the material is produced and the distance to the construction site. While we do not yet have data to definitively state that one system is always better than another, the differences between options within one construction type are often greater than the differences between construction types. Figure 2.5

shows the range of embodied carbon data for larger buildings of different structural materials. The range of embodied carbon for steel and concrete structures is high due to the range of system efficiency and material choices made.

Comparing LCA Methods

There are some industry reports on the environmental impacts of materials, such as those of the World Steel Association[6] or the National Ready Mix Concrete Association.[7] Most comprehensive databases are protected by intellectual property rights within commercial LCA software. The first open-source database of embodied energy and carbon, ICE,[8] was developed by the University of Bath (UK) some years ago. The Quartz Project Database[9] provides open source health and LCA data for select building materials. Neither of these are being updated. Development of EPD databases offer part of the solution. Other databases such as EcoInvent[10] and LCA software such as GaBi[11] enable users to customize data to reflect regional conditions yet require a license and LCA expertise to apply. Commercial and open-source software is available to perform WBLCA. For North America, the Impact Estimator for Buildings (IE4B) from Athena SMI[12] is available free of charge. Kieran Timberlake and Thinkstep released the Tally tool,[13] which extracts data from 3-D design models to calculate embodied impacts. A license is needed to use Tally. Governments in Europe have supported the development of regional LCA databases that integrate EPD results and standardize reporting, and

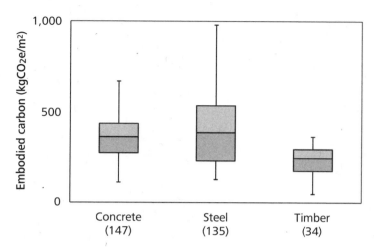

Fig. 2.5: *Ranges of embodied carbon for different structural material types from deQo database.* Credit: Adapted from C. De Wolf, *Low Carbon Pathways for Structural Design: Embodied Life Cycle Impacts of Building Structures*

a range of LCA tools have been developed globally.

Concrete

The ECC of concrete varies significantly depending on the strength, the cement content, the percentage of cement replaced by supplementary cementitious materials (SCMs), and the amount of reinforcing steel (rebar) used.

Figure 2.6 illustrates the variability between different data sources and different strengths of concrete. The variation can range from 0.08 to 0.22 kgCO2e/kg with the same assumptions on rebar percentage and cement replacements due to different strengths and databases.

Though cement only accounts for about 10 percent of the weight of typical modern concretes, it accounts for around 90 percent of the embodied carbon. When exposed to air, the outer inch or so of hardened concrete

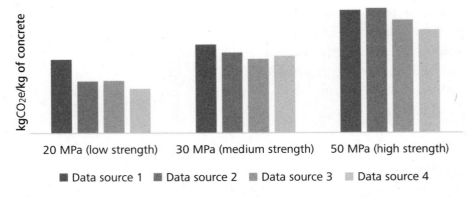

Fig. 2.6: *Varying measures of concrete's embodied carbon by strength and data source.*
Credit: Adapted from C. DeWolf, *Low Carbon Pathways*

absorbs CO_2 in a process called *carbonation*, thereby reducing concrete's carbon footprint, but this process is too minor and slow to have appreciable effect in embodied carbon calculations.[14]

Developments in lower-carbon concrete mixes are discussed in Chapter Six: Concrete.

Steel

The ECC of steel varies due to its *recycled content* (how much recycled steel is used to make new steel) and also *recycling rate* (how much steel might be recycled at the end of life), as well as the type of fuel used for its production, the availability of scrap steel,

transport distance and mode, and the end product. To date, steel ECC values also vary widely just based on data sources, as shown in Figure 2.7.

Two methods exist to produce steel: from virgin iron ore (primary steel) or from recycled steel scrap (secondary steel). The first method uses a Basic Oxygen Furnace (BOF) and requires about 25 MJ/kg, the second method, or Electric Arc Furnace (EAF), only requires 9 MJ/kg. The ECCs for steel, intimately tied to energy type and amount, are comparably different for primary and secondary steel. While 95 percent of structural steel and 70 percent of rebar is recycled, the *recycled content* of steel is considerably lower (at present) due to the shortage of scrap.

The ECC of steel is extremely dependent on the region where it is produced, as the results are sensitive to the emissions factor[15] of the energy mix (kgCO2e/kWh) as well as to the recycled content determined by the available scrap steel in the corresponding regions.[16, 17] With more accurate information on these key factors for a specific region where the supplying steel is produced, an adapted ECC can be found for structural steel and rebar. Scrap becoming more available in

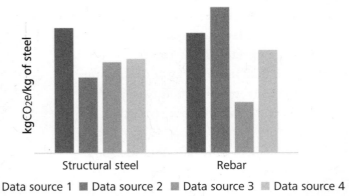

Fig. 2.7: *Varying measures of steel's embodied carbon by type and data source.* Credit: C. DeWolf, *Low Carbon Pathways*

the future will significantly reduce the ECC of steel by 2050 if the steel and engineering industry works toward energy and material efficiency. Currently available technologies only allow for five percent GHG emission reduction between 2014 and 2030.[18] Alternative steelmaking processes need to be developed in order to make more substantial reductions. Incremental carbon reductions can be obtained from heat recovery from blast furnace slag and from waste heat in electric arc furnaces, the use of by-products for the production of base chemicals, and the production of high-quality steel from scrap-based secondary steelmaking.

Wood

Wood is the most technologically advanced material I can build with. It just happens to be that Mother Nature holds the patent on it and we are not comfortable with that. But that's the way it should be: nature's fingerprints in the built environment.

— Michael Green, February 2013,
TED talk, Vancouver.

Wood does indeed grow on trees. But its ECC varies regionally due to the type of tree, the final wood products, proximity to construction markets, and variations in energy used to dry and manufacture final products. To dry lumber in a kiln, off-cuts are typically burnt to provide energy.

When calculating the ECC of wood, an important debate arises around *carbon sequestration*. During their lifetime, trees absorb carbon while growing; a ton of carbon in a tree is 3.67 tons of CO_2 removed from the air (see "A Word about carbon" in the Introduction). On that basis, the wood industry argues that a negative number should be used for the ECC of lumber to account for the carbon sequestered. However, the sequestration rate depends highly on the end-of-life treatment of the lumber product, the efficiency of manufacturing, and on the sustainability of the source forest management. In Chapter Four: Wood, Jason Grant talks more about sustainable forestry and sequestration rates.

There are two types of manufactured lumber: *sawn* or *milled* lumber, and *engineered lumber*. Sawn lumber has a low ECC as its impacts result primarily from harvesting and transport emissions. Engineered lumber has a higher ECC as it requires adhesives and processing. And there are two broad categories of wood: *softwood* comes from coniferous species such as pine, fir, spruce, or cedar, and tends to grow faster resulting in a lower density and strength, whereas *hardwood* comes from deciduous trees such as maple, oak, walnut, or alder and tends to be slower growing, stronger, and denser. Most structural sawn lumber is softwood. Engineered lumber is a composite material using mostly softwoods combined with glue to make the structural elements into idealized shapes such as sheets, beams, and I-shapes. Examples are plywood, cross-laminated timber (CLT), and glued-laminated lumber (glulam). The layers of CLT, a recent innovation, are glued perpendicular to adjacent layers giving strength in two directions, making it ideal for walls, floors, and roofs. (See the section on mass timber in Chapter Four.) The layers of glulam are glued in the same direction, making it ideal for columns, beams, and curved shapes. Forest products include not only structural shapes, but also pulp chips, sawdust, shavings, wood fiber, and bark, which can be used as fuel or sometimes in other building products like particleboard.

Two main scenarios for lumber's end-of-life treatment exist today: landfill and incineration for energy. Reuse is also possible for larger wood members, but is rare. Many argue that the carbon sequestered in lumber should only be accounted for *if the wood is known to be from a sustainable source* such as the Forest Stewardship Council (FSC). Jason Grant explains this in more detail in Chapter Four.

Figure 2.8 compares glulam, CLT, sawn hardwood, and sawn softwood. The reasons for the variations in the ECC of lumber are multiple. First, engineered, sawn, or whole lumber have different ECCs. Second, carbon sequestration and sustainable forest management assumptions vary from one study to another. Third, the end-of-life scenarios are unclear: landfill, reuse, recycling, use as fuel, etc. have different impacts. Fourth, the provenance and type of wood or forest are important. Fifth, the transport and availability of local lumber play a role. Moreover, the different data sources are applied in different regions, leading to varying assumptions in calculating the ECC of lumber.

Other Structural Materials

Though the main structural materials used in the industrialized building industry are concrete (and concrete block), steel, and lumber, many other lower-carbon alternatives exist. Rammed earth, cob, and adobe, for example, are old earthen building systems — and extremely low-carbon ones — now enjoying a worldwide renaissance. Chapter Six: Concrete gives some examples. Structural bamboo species as grown in the tropics produce very tough, strong fibers, and are far more productive per acre than any hardwoods or softwoods. With the advent of advanced steel fastening systems and various glulam shapes, bamboo is also enjoying a new life in construction. No less promising are the many new and old structural technologies appearing in architecture today, such as Shigeru Ban's cardboard columns, load-bearing straw bales, fabric structures, Guastavino's historic brick vaults, and more. As the rules and supporting data for LCA mature, designers will

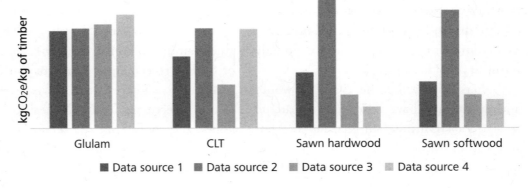

Fig. 2.8: *Varying measures of wood's embodied carbon by type and data source.* Credit: C. DeWolf, *Low Carbon Pathways*

kgCO₂e/kg of timber

Glulam   CLT   Sawn hardwood   Sawn softwood

■ Data source 1   ■ Data source 2   ■ Data source 3   ■ Data source 4

be able to objectively evaluate each for its environmental impacts as well as functional qualities.

Nonstructural Materials

As stated, structural materials comprise about half of the embodied carbon footprint of most projects, making them the primary focus of an effort to reduce embodied carbon. As Larry Strain shows in Chapter Three: Rebuild, the remaining embodied carbon emissions of a building are dispersed among many components.

One category of nonstructural materials does stand out, however, in part because of its popularity in net-zero energy and super-insulated buildings: spray foam insulation. The first generation of blowing agents for spray polyurethane, polystyrene, and extruded polystyrene foams had climate impacts, or global warming potential (GWP), sometimes thousands of times as much per pound as CO_2; their use was a benefit to the building occupants and owner, but bad news for the climate. The industry is now on fourth and fifth generations of blowing agents that are substantially closer to CO_2 in their GWP (sometimes they *are* CO_2!), but foam insulations in general still have a relatively high ECC compared to other insulation options.

We haven't tried to determine which structural material has the lowest environmental impact, but to offer strategies for calculating the embodied carbon of materials and to illustrate what the challenges are. Regional variability of ECCs is high. GWP of materials is only one impact factor among others including toxicity, resource depletion, loss of biodiversity, etc. Many argue that the embodied carbon of materials per unit volume or mass should not be used (solely) to make decisions for minimizing the harmful effects of construction, but to perform calculations at the structural, building, and city scale. Others contend that the long-term effect of carbon emissions dwarfs other concerns. Regardless, at present we can only make rough calculations. But the science is rapidly growing and maturing, and as transparent, reliable values for the ECC of typical building materials become available, designers will be able to make informed decisions about the lowest carbon options for their projects. Consideration for the carbon footprint and toxicity of construction materials should soon be as fundamental as designing for fire, wind, and earthquake.

Comparing the Embodied Carbon of Buildings

When looking to understand and reduce the embodied impacts of a building, we naturally ask, "What design choices make for a low carbon building?" The short answer is to use lower-carbon or better carbon-sequestering materials, and use less material overall. Using less material means: build smaller buildings, configure buildings to enable efficient structural systems and/or eliminate additional finish materials, and, especially, reuse existing buildings and building materials.

The Embodied Carbon Benchmark Project (ECB) is a compilation of whole building LCA studies to discern the order of magnitude and range of whole building embodied carbon estimates.[19] This database records the initial embodied carbon (cradle-to-site) per unit area, with note of building use, number of stories, and floor area. With over a thousand buildings in the database, this is currently the largest known collection of whole building embodied carbon studies. We found that over 95 percent of the studies reported the initial embodied carbon to be less than 1,000 $kgCO_2e/m^2$ and nearly 75 percent of office buildings (the most populated building type) have embodied impacts less than 500 $kg\,CO_2e/m^2$. Figure 2.9 shows a summary of the data without any outliers removed. Note that there are buildings with embodied carbon more than four times the average value and others with embodied carbon of less than 25 percent of the average.

There are limitations to ECB data, as it generally does not include maintenance, energy used to operate the building (lighting and heating etc.), or end-of-life impacts, nor building-related components such as site work, mechanical/electrical systems, and furnishings. Further, while most building LCAs include foundations, superstructure, and sometimes enclosure, some also include finishes while most do not. In short, the ECB database is unavoidably a bit of an incomplete and imperfect collection of data, but a no less necessary first step in discerning the embodied carbon footprint of buildings.

In analyzing the data, we looked to identify trends such as: Do taller and/or larger buildings have larger embodied carbon impacts? Do some building uses tend to have larger impacts? In Figure 2.9, the embodied carbon is plotted based upon number of stories, demonstrating that buildings at all

scales can be relatively low or high carbon. Of note, some projects report data including a negative value for the carbon sequestered in wood products, contributing to the lower numbers shown. As mentioned, the method for accounting for this sequestered "biogenic" carbon is not uniformly agreed upon.

One of the key findings of the ECB project is that we urgently need to standardize the embodied carbon calculation and reporting methods — that is, LCA. Databases such as the *database of embodied Quantity outputs* (deQo)[20] that collects structural material quantities in a standardized manner can reach more detailed conclusions such as identifying a trend of increasing materials usage for larger/taller buildings as shown in Figure 2.10.

In deQo, data on material quantities and embodied carbon in buildings is collected in close collaboration with leading structural design firms worldwide. There is no consensus for evaluating embodied carbon in current practices, but deQo uses a rigorous and transparent methodology to assess the embodied carbon of building structures. DeQo collects material quantities; once updated ECCs are available, they can be applied for more accurate results for the embodied carbon of a building as a whole.

Getting to Zero: Embodied Carbon

We know that the embodied carbon of buildings is significant, both for its magnitude and for the time-value effect described in Chapter One. However, these are the early days of a nascent science; we may have accuracy but not great precision. We are still in

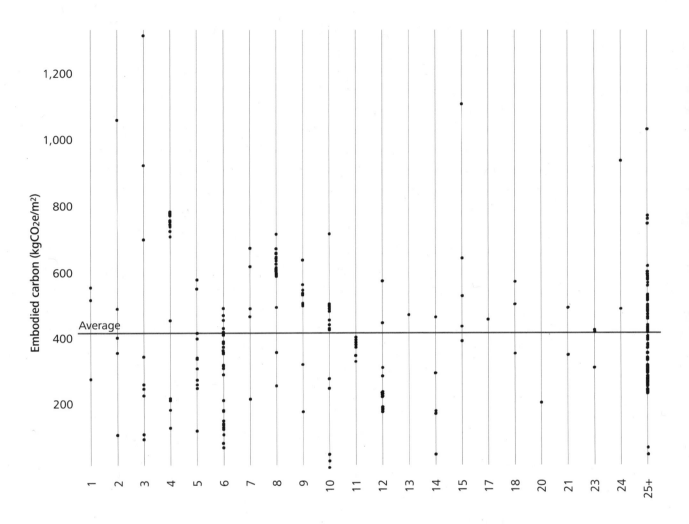

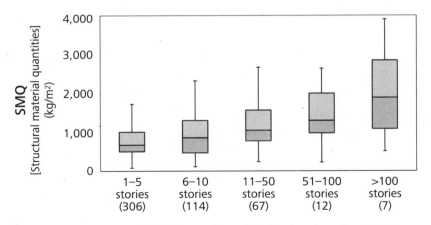

Fig. 2.9, above: *Embodied carbon for life cycle stage A according to building stories.* Credit: Adapted from K. Simonen et al., *Embodied Carbon Benchmark Study: LCA for Low Carbon Construction*

Fig. 2.10, left: *Material quantities increasing with height extracted from deQo.* Credit: Adapted from C. De Wolf, *Low Carbon Pathways*

the early stages of measuring, cataloging, and analyzing buildings of different types and in different regions to identify exact embodied carbon benchmarks or targets by which to design and regulate the built environment. Current regional initiatives in green building rating systems are rewarding (or mandating) the use of LCA to track and refine embodied carbon benchmarks. A group of structural engineers in the US is working with the Carbon Leadership Forum to establish a Structural Engineers 2050 Initiative aimed at motivating engineers to track and report structural material quantities so as to aid our understanding of the range of structural material climate impacts.

In order to meet targets set in the Paris Climate Agreement, we will need to be globally carbon neutral by 2050, and that includes the building material and product industry. We already have many buildings demonstrating net-zero operational energy, yet getting to zero embodied carbon can seem impossible.

In fact it *is* impossible, let's get real. Even a simple adobe hut has a few wires and windows (embodied carbon), and that wood-burning stove in the corner is effective and romantic but nonetheless emits operational carbon. Buildings and cars and smartphones have impacts, let's not pretend otherwise. That said, it may turn out that our impossible goal will one day soon be easy: the collective reports in this book point to a new carbon architecture that literally absorbs more carbon in its construction than it emits. Pull carbon out of the sky and turn it into buildings. Read on to see how it's already happening.

Notes

1. Ochsendorf, J., et al. 2011. *Methods, Impacts, and Opportunities in the Concrete Building Life Cycle, Research Report R11-01*. Concrete Sustainability Hub, Department of Civil and Environmental Engineering. Cambridge, MA. Massachusetts Institute of Technology.

2. Simonen, K. 2014. "Life Cycle Assessment: Pocket Architecture Technical Design Series." Routledge, London, UK.

3. Simonen, K., et al. 2015, Integrating Environmental Impacts as Another Measure of Earthquake Performance for Tall Buildings in High Seismic Zones. Structures Congress 2015: 933–944.

4. Giesekam, J., Barrett, J. R., & Taylor, P. 2016. "Construction sector views on low carbon building materials." *Building Research & Information*, 44(4), 423–444.

5. Kaethner, S., and Burridge, J. 2012. "Embodied CO_2 of structural frames." *The Structural Engineer*. May, 33–40.

6. World Steel Association. 2016. *Life cycle thinking*. www.worldsteel.org/steel-by-topic/life-cycle-assessment.html

7. National Ready Mixed Concrete Association (NRMCA). 2016. *Sustainability and the Concrete Industry*. www.nrmca.org/sustainability/index.asp.

8. Hammond, G., and C. Jones. 2010. *Inventory of Carbon and Energy (ICE), Version 1.6a*. Sustainable Energy Research Team (SERT), Department of Mechanical Engineering. Bath, UK: University of Bath.

9. Quartz 2016. Quartz database for common building products: [Quartz_db_2016_Dec.xlsx]. Retrieved fromquartzproject.org.

10. EcoInvent, Swiss Centre for Life Cycle Inventories, Retrieved May 13, 2016, from www.ecoinvent.ch.

11. GaBi PE International. 2016. *GaBi 4 extension database III: Steel module and GaBi 4 extension database XIV: construction materials module.* Retrieved December 1, 2016, from www.gabi-software.com

12. Athena SMI. 2009. Impact Estimator for Buildings. Available fromathenasmi.org/our-software-data/overview/.

13. Tally. 2016. Revit add-in Tally™. Real Time Environmental Impact tool. Revised Tally™ Revit Application, Kieran Timberlake Research Group, Retrieved February 2017, from www.kierantimberlake.com/pages/view/95/tally/parent:4

14. Webster, M. D., et al. 2012. *Structure and Carbon — How Materials Affect the Climate.* SEI Sustainability Committee, Carbon Working Group. Reston, VA: American Society of Civil Engineers (ASCE).

15. IEA. 2016. *CO$_2$ Emissions.* International Energy Agency, Retrieved October 22, 2016, from www.iea.org/statistics/topics/co2emis sions/

16. Wübbeke, J., and Heroth, T. 2014. "Challenges and political solutions for steel recycling in China." *Resources, Conservation and Recycling,* 87, 1–7.

17. EurActiv. 2016. *Steel recycling on the rise.* Retrieved December 28, 2016, from www.euractiv.com/section/sustainable-dev/news/steel-recycling-on-the-rise

18. Arens, M., et al. 2016. "Pathways to a low-carbon iron and steel industry in the medium-term: The case of Germany." *Journal of Cleaner Production,* 1–15, doi: 10.1016/j.jclepro.2015.12.097.

19. Simonen, K., B. Rodriguez, and S. Li. 2017. *Embodied Carbon Benchmark Data Visualization,* database/website, www.carbon leadershipforum.org/data-visualization/

20. Database of embodied Quantity outputs (deQo), deqo.mit.edu

Chapter Three

Rebuild: What You Build Matters, What You Don't Build Matters More

by Larry Strain

NEW BUILDINGS ARE A PROBLEM. The current gold standard for reducing emissions from buildings is to build new, net-zero energy (NZE) buildings — very efficient and powered by renewable energy sources, where the energy generated is equal to or more than the energy needed to operate them. Given that we will always need new buildings, this is a critical piece of getting to a carbon neutral built environment. But there

Editor's Introduction

In the ten-thousand-year history of building, people always built with what was nearby in the landscape, and without rearranging materials too much if at all. The only exceptions were people with access to unusual amounts of energy: in ancient times, the very powerful who had access to human energy (mostly slaves), and in the last 200 years, the huge proportion of humanity with access to fossil fuel energy. If we want to reduce the carbon footprint of our buildings, then we have to think the same way: don't move things very far, don't rearrange molecules very much, and use what you already have. In the countryside, and throughout most of history in most places, that means using wood, clay, stone, straw, bamboo, and other so-called natural materials. In the cities — where most of us now live — the predominant resource at hand is — wait for it — existing buildings! Here Larry Strain makes the case for the enormous importance of preserving and upgrading existing building stock, especially when the alternative is to tear something down and replace it with a brand new structure. The carbon analysis can be maddeningly complex and often imprecise, but the conclusion is crystal clear: Don't tear it down — give it a facelift and an energy upgrade. That will usually be less expensive to the owner and less harmful to the climate.

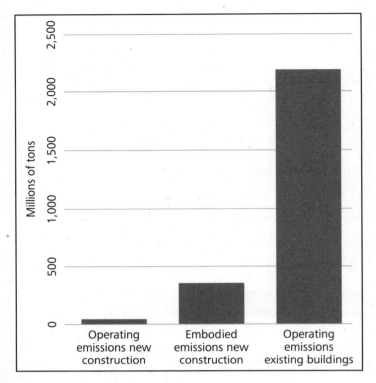

Fig. 3.1: *Annual carbon emissions: 6 billion square feet of new construction / 310 billion square feet of existing buildings.* Credit: Siegel & Strain Architects

fast enough. In the first place, we don't build enough — currently about six billion square feet per year — to make a difference. Operating six billion square feet of efficient, code-compliant buildings generates about 40 million tons of GHGs, less than one percent of total US emissions. Building six billion square feet will generate about 350 million tons, just over five percent of our annual emissions — a significant number, but it doesn't begin to compare with the 2.3 *billion* tons of emissions from operating our existing buildings, more than a third of US annual emissions.

Reuse: A Complete Strategy

Reusing buildings is one of the most effective and radical strategies we have for reducing carbon emissions from the built environment. Effective, because reusing and renovating a structure has a much lower embodied carbon footprint than building a new one, and if the renovation includes efficiency upgrades, it can also reduce the existing building's operating emissions. And radical, because being frugal and careful with our resources, such as our existing building stock, is not usually a priority. We like new, shiny things. We don't replace buildings because they wear out, we replace them because the land they are built on becomes more valuable, or we no longer like the way they look, or we just want something new. We also like new buildings because building anew is often easier than renovating existing structures. But renovations + upgrades usually save money, and they almost always save carbon.

is a problem with this strategy — building all of those new structures will generate a lot of emissions. Even though new NZE buildings don't add operating emissions, they still have a huge and immediate carbon footprint.

We Can't Build Our Way Out of This

In the US buildings are responsible for nearly half of greenhouse gas (GHG) emissions. Most of that is from operating our buildings, and a smaller but significant amount is from building our buildings, also known as embodied carbon.

New, efficient, super-green, even net-zero buildings won't reduce our emissions

In the following chapters, you're going to learn about ways to lower and even eliminate a building's embodied carbon footprint by using materials and construction methods that can lower and even sequester carbon. Most of these materials we already know about, some need to be developed, and some just need to get to market. But most of them require a shift in the way we currently build in this country — and that shift may take more time than we really have.

A strategy that we already understand, and that would significantly reduce carbon emissions, is to reuse and upgrade existing buildings and build fewer new ones. The embodied emissions from an existing structure have already happened. We can't get them back, but neither can we afford to release that amount again by replacing old buildings with new ones. Renovating a building releases somewhere between 50 to 75 percent less carbon than building a new one does. That's because most renovation projects don't replace the structure or the exterior envelope that contain most of the embodied carbon.

Perhaps a more compelling reason to reuse buildings is that it comes with an opportunity to address an even bigger source of GHG emissions: operating all the buildings we already have. When building renovations also include deep energy efficiency upgrades and we power those renovated buildings with renewable clean energy, we can reduce two sources of emissions at the same time — *embodied and operating.*

Reducing Embodied Carbon

Let's start with embodied carbon. If we want to reduce embodied carbon, it helps to know where it is — mostly in the materials that we build with. Construction equipment, transporting workers and materials to the jobsite, and site work also contribute emissions — but not as much. For remote sites, transport can be significant, and for large sites, site work emissions can be a larger percentage — but typically the majority is from materials.

It's also useful to know the carbon footprint for different types of buildings and to understand how the materials and their carbon emissions are distributed (Figure 3.3). On a square-foot basis, big heavy buildings have a much higher carbon footprint than small light buildings. It's partly because they weigh more but mostly because of what they are made of. Beyond a certain size, buildings usually have steel and concrete structural systems (although wood is now a viable alternative for large buildings, as you will read about later). Small light buildings, at least

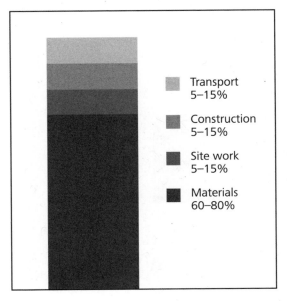

Fig. 3.2: *Where's the carbon?* Credit: Siegel & Strain Architects

in North America, have traditionally been framed in wood, which has a much lower carbon footprint than steel and concrete. So we want to avoid building more large heavy steel and concrete structures; these are the buildings we should be reusing and renovating. Again, it's worth noting that renovating them has an even lower carbon footprint than building small light new buildings.

As noted, renovation projects have lower embodied emissions than new construction because they typically reuse most of the structure and building envelope. But even renovation projects still have embodied emissions, and we can reduce those even further if we pay attention. For example, renovation projects often need to remove materials such as lay-in acoustic ceilings or worn-out carpet. Instead of replacing them, we may be able to use the underlying structure as the new interior finish and reduce emissions

and transform the space in the process. We can also use some of the low-carbon materials described in the following chapters in new creative ways — wrapping uninsulated metal warehouses in straw bales, using salvaged materials instead of new materials, or just replacing synthetic carpet with natural fiber carpets. We can also design new buildings and renovations knowing they will be renovated in the future, using building components that are removable, cleanable, and able to be refurbished. If people are able to change their buildings more easily, they may not be as likely to replace them.

Buildings aren't all we need to reuse; we also can reuse materials, and renovations generate a lot of "waste" materials. While new construction generates three to five pounds per square foot of solid waste, renovation projects can generate 20 to 30 times that much. If we can reuse those "waste" materials instead of discarding them, we save carbon.

Reducing Operating Carbon: Renovation + Upgrade

Compared to building a new building, renovating an existing building clearly saves embodied carbon emissions. But to get the most out of reuse, we also need to lower operating emissions. Energy upgrades are what sets reuse apart as a strategy for reducing total carbon emissions. We have a lot of existing buildings. They aren't as efficient as new code-compliant buildings, and they definitely are not net-zero, so if we don't reduce their operating emissions, we are missing a huge opportunity. And it's not just about making them a little more efficient, we

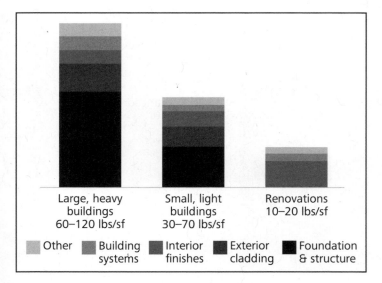

Fig. 3.3: *Carbon emissions by building type and material.*

Credit: Siegel & Strain Architects

should aim to make existing buildings substantially better if not net-zero.

Upgrading to Zero

There are three basic parts to making an existing building net-zero:

+ Improve systems efficiency: upgrade the lighting, HVAC systems, equipment, controls, etc.
+ Improve building efficiency: insulation, windows, shading, air sealing, daylighting
+ Power the building with renewable energy

For many structures, NZE remodels are not as hard as we might think.

Siegel & Strain Architects co-authored a study with Integral Group of a two-story office remodel and upgrade for DPR Construction. It was designed as a net-zero building and is currently generating more energy that it consumes (www.ecobuildnetwork.org/projects/total-carbon-study). This interior remodel upgraded equipment and lighting, added skylights and photovoltaic power (PVs), and made only minimal upgrades to building efficiency (roof insulation). The remodel generated about one third of the embodied emissions that a new building would have, and because it is producing more power than it uses, it is also paying off that embodied carbon debt.

It wasn't even an ideal candidate for a net-zero retrofit. It is partly shaded by taller buildings, and the single-glazed aluminum storefront windows couldn't be replaced because they were historic. The really compelling part of this story is that even without ideal conditions, it made sense to retrofit; the project came in on budget and on time.

The priorities for deep energy upgrades for existing buildings have changed over the last ten years. It used to be that you always started by making the building more efficient — more insulation, new high-performance windows, new efficient equipment and lighting — and after you made it as efficient as possible, then maybe you added renewables to power the building. This was

Fig. 3.4: *DPR office remodel: rooftop solar arrays.*
Credit: Ted van der Linden, courtesy of DPR Construction

mostly because photovoltaic panels cost a lot, so you wanted to buy as few as possible. With the cost of PVs dropping and the advent of new efficient and relatively inexpensive heat pump technology, upgrading the equipment and adding PVs may be among the first things we do, not the last. Efficiency upgrades are still the place to start, but technology and economics have changed the equation.

Efficiency strategies also vary depending on whether the building is residential or commercial. Commercial buildings are often internally loaded, which means the heating and cooling loads are driven more by the lighting, equipment, and people in the building than by outside temperature (although heating and cooling loads can be high for glass skyscrapers). For commercial buildings, lighting and equipment upgrades will typically have the biggest impact on reducing energy and emissions. For residential buildings, which are dominated by heating and cooling, envelope upgrades will have a bigger impact, although appliances and equipment upgrades are also important.

The other thing that has changed is the urgency of climate change. We need to consider the total carbon emissions over the next 10 to 20 years and evaluate the initial carbon investment for energy efficiency strategies against the future savings generated from those upgrades. How much carbon did we spend to reduce operating emissions, and how long will it take the savings from increased efficiency and clean energy production to offset that initial investment? When you do this analysis, it may change your approach to efficiency upgrades. Blowing in insulation,

recommissioning or even replacing inefficient HVAC and lighting systems are likely to have a good return on carbon invested; reskinning a building with a high-performance aluminum/glass curtain wall or wrapping the building in foam insulation may not be a good investment from a carbon standpoint. We need carbon reduction strategies that have a positive payback within a 10- to 15-year time frame, or we should look for other strategies.

Retrofit Opportunities

There are two reasons to renovate and upgrade existing buildings. The first is to reduce operating emissions from existing buildings, and that applies to all buildings. The second is to reduce embodied emissions by renovating existing structures instead of building new ones. This strategy makes the most sense for larger commercial buildings because they have a higher embodied carbon footprint than smaller residential ones. The good news is we have a lot of these buildings. There are almost six million commercial buildings in the US, and the majority of them are one-to-three-story, flat-roofed office buildings. We also have a plethora of one-story strip malls, warehouses, and schools, also with large expanses of flat roofs. These are prime candidates for efficiency and net-zero upgrades.

Energy Efficiency Opportunities

We need to identify the best candidates for energy upgrades. Start with the buildings that use a lot of energy compared to similar buildings with similar uses. Poor-performing buildings have a higher potential to save energy and reduce carbon emissions than more efficient buildings. These build-ings usually have:

+ poor thermal envelopes: little or no insu-lation, single-glazed windows, unshaded windows, leaky drafty buildings
+ old, inefficient HVAC and lighting systems and controls, equipment, and appliances

Net-zero Opportunities

We need to identify the best candidates for zero net energy upgrades.

+ buildings with unshaded flat roofs or south- (in the northern hemisphere) and west-facing sloped roofs

+ 1–3 story buildings
+ buildings with adjacent unshaded land: parking lots with PV canopies produce power and have the added benefit of shad-ing the cars and pavement and reducing the heat island effect around the building

Saving Embodied Carbon Opportunities

We need to identify the best candidates for saving embodied carbon.

+ reuse high eCO_2 buildings (heavy steel and concrete buildings) instead of building new steel and concrete buildings

Chapter Four

Wood: Like Never Before

Mass Timber Construction

by Frances Yang and Andrew Lawrence

THE STRUCTURAL PALETTE of materials for large- and medium-rise construction until recently consisted solely of concrete and steel, which can often be quite energy and carbon intensive. Furthermore, with the globalization of materials markets and greater transport costs, the environmental impacts of all construction materials grows.

Expanding the palette with wood, a renewable and carbon-sequestering material, could help the climate and provide benefit to rural forest areas. To do this, we need to take a two-pronged approach that ensures carbon savings while targeting the current barriers to going taller with wood.

Fig. 4.1: *National Assembly for Wales.*
Credit: Redshift Photography, courtesy of Arup

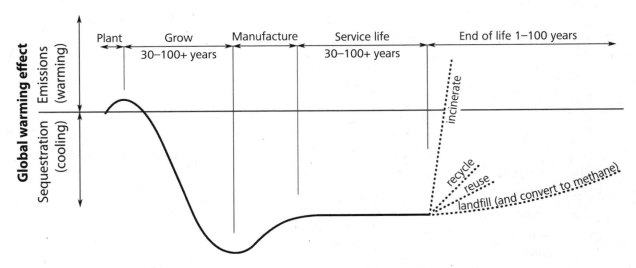

Fig. 4.2: *Depiction of carbon emissions over the life of a wood product.* Credit: Arup / Bruce King

Fig. 4.3: *Framework in Portland, Oregon: a winner of the USDA Tall Wood Demonstration Prize.* Credit: LEVER Architects

The Carbon Argument

Wood sequesters carbon during the growth of trees, which is illustrated by the steep drop down in CO_2 emissions on the far left side of the curve in Figure 4.2 above. Note that the CO_2 emissions typical during planting (from equipment and fertilizers), manufacturing, and use stages is minor in comparison, and only at end of life is there a chance that the emissions might return above the zero line, depending on what becomes of the wood after its first use. If one likens wood to a "CO_2 sponge," the amount of CO_2 released at end of life depends on how you "squeeze it out."

Arup, an international engineering firm committed to sustainable development, used the figure above as the basis for the embodied carbon studies on a 12-story mixed-used project in Portland, Oregon, to explain the scale of difference between the wood solution and a conventional steel/concrete structure (see Figure 4.3). This building uses cross-laminated timber (CLT) and glued-laminated (glulam) members as the primary frame instead of flat-slab concrete, as would be the more conventional system. The study not only shows the potential carbon savings compared to conventional concrete and steel structures, it also provides direction for setting priorities.

Life cycle assessment (LCA) is a scientific method to quantifiably estimate the environmental impacts of products and processes, and Arup felt that the project team for this building should integrate LCA into the design process from the beginning. LCA has the most potential impact when used not merely to record what has been done at the end, but also to highlight carbon-saving opportunities as early as possible, before those chances are shut out in later design phases. Building-level assessments are also useful for identifying any trade-offs between embodied vs. operational emissions to arrive at an optimal, integrated solution. While LCA accounts for many environmental impacts, global warming potential (GWP) is the indicator that tracks embodied carbon.

The embodied carbon results for the alternative designs presented in the charts below (Figures 4.4 and 4.5) show the project-specific sources of impacts by life cycle phase. The difference in the end points shows the overall reduction that can be anticipated based on currently established supply chains and practices, including a worst-case sourcing from furthest transport distances by road, and the current practices of demolition and disposal at end of life. The factors are from the Athena Impact Estimator for Buildings v5.1, coupled with a carbon sequestration calculator by FP Innovations and the Athena Sustainable Materials Institute.

The comparison is based on preliminary material quantities for the primary

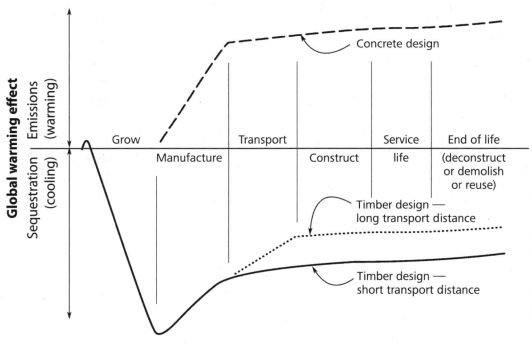

Fig. 4.4: *Comparison of carbon emissions between timber design and concrete design over life cycle of structural materials in 12-story tower.* Credit: Arup / Bruce King

structural system, including foundations, from the structural engineers. It does not include nonstructural materials common to both the concrete and wood systems such as façade, finishes, and services. In order to maintain a like-for-like comparison, it does include additional materials unique to either system when required to achieve the same functional unit on the basis of acoustic, aesthetic, fire, and durability performance, in addition to structural performance.

These results confirmed the sequestration potential of using wood instead of steel or concrete to reduce carbon emissions. It was also found that sourcing materials locally reduced transportation emissions, but those savings paled in comparison to the CO_2 sequestered earlier on.

The team then looked at results based on other wood sources. Figure 4.5 shows the hypothetical case of sourcing wood that is not from a sustainably managed forest (by which carbon storage levels can be guaranteed). Of most surprise to the project team was that the final life cycle emissions (the end of each line) of the timber design could arrive very close to the final emissions of the concrete design option if both worst-case sourcing and worst-case transport scenarios were realized.

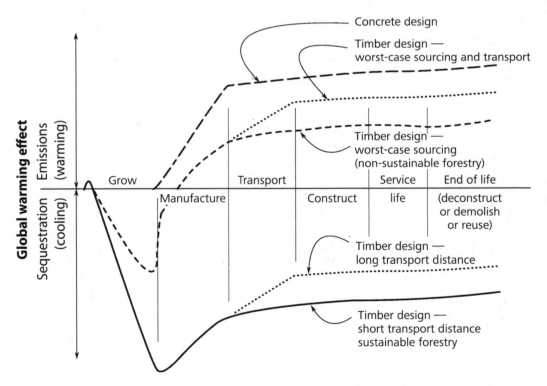

Fig. 4.5: *Comparison of carbon emissions between timber design and concrete design over life cycle of structural materials in 12-story tower, when wood is not sourced from sustainably managed forests.*
Credit: Arup / Bruce King

This further convinced the project team that selecting wood from sustainably managed forests was essential to the integrity of this project's sustainability claims. (See also Jason Grant's section on sustainable forestry in this chapter.) Forest conservation and careful management help to ensure our forests and other lands continue to remove carbon from the atmosphere while also improving soil and water quality, reducing wildfire risk, and otherwise helping forests to be more resilient in the face of climate change. United States and Canadian forests are governed by policies that demand, track, and report a stable carbon pool, typically through state and provincial programs. Still, these programs rely on a multitude of individual caretakers who may have different management methods on their sometimes small patch of land within the larger area that is measured. Thus, third-party certification programs using chains of custody are often regarded as the most robust mechanism to link sustainable harvest practices to the procured forest products, although they do carry high costs that can be a barrier to use for very small forest owners.

While not shown in the figures, the project team also examined the GWP emissions reported in LCA tools for the Beyond Building Life phase, and found a surprising amount, due to current practices of landfilling most construction wood waste. This led the team to question what the project could do to reduce emissions at that stage. In other building product industries, emissions from landfills are best addressed through policies that keep waste from landfills (where wood rots to emit methane) and programs that incentivize recovery of products at the end of their service in a building. Unfortunately, the lack of any kind of take-back program due to the very premature state of CLT manufacturing in the US, as well as the distant occurrence of these emissions for structural products (as compared to, for example, carpet and ceiling tiles), made tackling this phenomenon very challenging. One tactic, only barely used as yet, is to require suppliers of fabricated wood products (as with steel products in some countries) to clearly label each product with its material properties and/or governing standard, thus facilitating future reuse. In any case, mass timber is less likely to be thrown away, due to its higher value and greater technical potential for reuse. Reusing or repurposing mass timber products as other construction products would then reduce the embodied carbon footprint of those future buildings.

So How Tall Can Timber Really Go?

Tall wood buildings are not entirely new, and some examples built in the past centuries can be found in the USA, Japan, and other places. But those used massive pieces from old-growth trees, which is no longer feasible or desirable, and in the past hundred years, tall buildings have been predominantly built of steel and concrete. In the last 15 years, we have seen a renaissance of timber buildings growing from modest code-limited three- to four-story townhouses to 10+ story towers, thanks to technical advances and changes in regulations. The 18-story Brock Commons building, recently completed in Vancouver, relies on concrete cores to resist wind loading, while the 19-story Haut tower in Amsterdam (see Figure 4.6), now being designed by Arup and Team V Architectuur, will rely partly on timber walls to resist the force of the wind. So how tall can timber really go and what are the limitations?

To understand that, we need to understand that there are many different ways of building with wood. The three- to four-story townhouses of the 1990s were lightweight timber frames built from thin studs, joists, and plywood. Also known as stick framing,

Fig. 4.6: *Haut tower in Amsterdam.* Credit: Team V Architectuur

this is an efficient and economic form of construction, but also relatively weak and therefore quite limited in height. Wind and earthquake are resisted by the large number of walls throughout the building, and it's well-suited for the typical one- and two-story North American house for which it was invented back in the 1830s. In Europe the studs, joists, and plywood are often prefabricated off-site into floor and wall "cassettes."

Enter Cross-laminated Timber (CLT)

CLT was invented in 1998 and offers a much stronger form of construction using solid timber panels rather than lightweight cassettes. With CLT, boards are laminated in alternating directions (which is why some people call it "jumbo plywood"), creating panels that are stronger and stiffer than stick-framed systems. CLT panels are considered to be heavy or "mass" timber, meaning they have some inherent fire resistance and the ability to be used confidently at scale. This enabled Waugh Thistleton's Murray Grove project in London in 2008. This was essentially an eight-story CLT tube perforated by small perimeter windows. This was followed in 2012 by Lend Lease's nine-story Forte tower in Melbourne, again entirely in CLT.

CLT systems are prefabricated, such that a building becomes rather like a giant piece of flat-pack furniture. The kit-of-parts nature of the system also lends itself to disassembly and easier reuse in the future.

A large added bonus is that we can utilize available timber that no longer offers sequestration benefits, such as beetle-killed pine, still plentiful in the western US. Use of "dead" wood (as long as it is felled before it has decayed) has multiple benefits: it clears the forest of fire hazard, upcycles the material into a functional resource, and gives the local economy a feedstock from which they can create jobs and profit by adding value. A blue tinge can sometimes be seen in the CLT panels made from beetle-kill pine, a telltale sign of beetle-kill in the wood's past.

Stiffness

In designing Haut, Arup and Team V sought to push timber to its limits. This meant a fundamental change from the previous CLT buildings that had been built as *platform frames* (Figure 4.7). In a platform frame, each successive floor sits on the walls below; this is easy to construct, but as the wind blows, the CLT floors (sandwiched between the walls above and below) are compressed. Wood across the grain is relatively soft, so this compression of the floors greatly reduces the lateral stiffness of the building. Rigidity can be improved with grout packers or by castellation of the wall/slab connection, but to maximize the stiffness, the CLT wall panels need to be built directly off the wall panels below, with the floors supported on brackets.

Lateral stiffness under wind loading is a vital consideration, because tall building design is often governed by occupants' comfort in strong winds. This stiffness depends partly on the damping or internal friction within the structure, which dictates how quickly the wind vibrations will die away. However, so few tall timber structures have been built

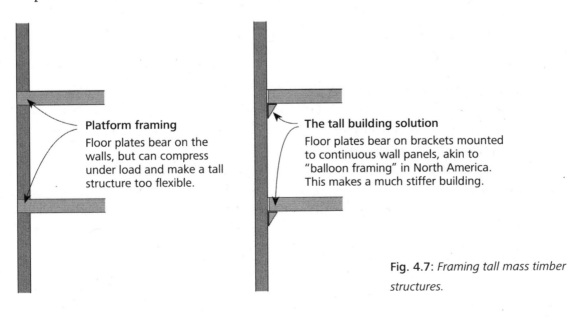

Platform framing
Floor plates bear on the walls, but can compress under load and make a tall structure too flexible.

The tall building solution
Floor plates bear on brackets mounted to continuous wall panels, akin to "balloon framing" in North America. This makes a much stiffer building.

Fig. 4.7: *Framing tall mass timber structures.*

to date that there is little data on the amount of damping inherent in a CLT building. That is why it's so important to increase the height of timber buildings progressively, monitor the vibration of the buildings, and then share that data across the industry.

Fire

Fire engineering of tall timber buildings is a challenge that needs to be carefully addressed. Most of the taller timber structures built so far have relied on *encapsulation* — protecting the wood from the heat of the fire using gypsum board. This extra layer adds construction time and cost, and also hides the beauty of the wooden structure.

CLT structures' classification as "heavy timber" means they are thick enough to retain at least some strength even when fire chars the outer inch or so. However, the risk of leaving the wood exposed (without encapsulation) is that it could continue burning even after burnout of the building contents. This would obviously not be desirable, particularly in multi-occupancy residential structures where occupants do not always evacuate immediately. In a dense urban environment, it also presents the risk of the fire spreading to other nearby buildings. Sprinklers are very useful, but we cannot always fully rely on them, and building codes often require consideration of the (unlikely) scenario of the sprinklers failing.

To address these risks and also to understand the potential strengths of timber in fire, Arup collaborated with the University of Edinburgh to conduct full-scale fire tests on CLT compartments. (The research was also supported by KLH manufacturers of CLT, the UK Engineering and Physical Sciences Research Council Impact Acceleration and Schraubenwerk Gaisbach, GmbH screw manufacturers). The results show that leaving at least some of the wood exposed might now be possible, relying on the fact that wood will only continue burning if there is a sufficient source of external heat. (In the same way, a single log on a fire will self-extinguish unless there are enough hot embers to keep it warm.) For the Framework building in Portland, winner of the US Tall Wood Building Prize, fire testing has informed the analysis to prove that timber columns, beams, and areas of the CLT ceiling soffit can be left exposed. Of course, even if the wood self-extinguishes, there is still a practical cost limit, since the longer fire resistance periods needed for taller buildings demand thicker cross-sections to survive charring.

Acoustics

For concrete and steel-framed buildings, the rules that govern acoustic performance, both airborne (such as speech) and impact (such as footsteps), of the internal separating constructions are relatively well understood, established, and accepted, but the same cannot be said for mass timber structures — irrespective of whether they are considered to be tall or otherwise.

The acoustic performance of timber constructions is a particularly complex design issue, not fully predictable based on the current levels of knowledge. The current approach utilized by acoustic designers is based largely on traditional analytic acoustics and semi-empirical methodologies. This can as-sess acoustic performance, at least within an order of magnitude, and the issues which are relevant to particular construction forms. However, unless this approach includes an acceptable level of safety between the es-timated performance and that required, laboratory and/or mock-up testing can often be required in order to accurately determine the performance of candidate constructions (including flanking elements and critical in-terface detailing).

Research currently ongoing in several universities is aimed toward developing analytical tools to predict the acoustic per-formance of timber constructions, though the time frame in which this might develop is uncertain.

Seismic Performance

There are not yet any codified mass timber seismic systems for the engineer to apply, so solutions are based on research and testing. Various systems — both hybrid and timber-only — have been developed for seismic areas and are finding application in various current projects. A typical hybrid solution is to have a conventional concrete core shear wall system to provide the necessary lateral strength and seismic resistance accompany-ing a timber-framed gravity system. Simply using plywood or CLT panels as shear walls does not work for taller structures because their greater flexibility, compared to concrete and steel solutions, leads to too much deflec-tion at the top of the structure. Two novel timber-only solutions are: (1) using glulam beams and columns to form braced frames,

and (2) post-tensioning CLT panels at the two ends of the wall, creating a rubber-band effect that pulls the building back to plumb. Research, testing, and analytical models of such solutions are showing that these build-ings perform to code requirements, and as their project-specific development continues, they will be added to the engineer's solution tool box. Non-codified solutions require an extensive testing and approvals process that adds cost and prolongs project schedules, but the work currently in progress will help the spread of taller wood buildings by add-ing to our knowledge.

Beyond Carbon

As we work to address the barriers to build-ing taller with wood, we need to consider more than the carbon benefits. First, carbon accounting based on LCA has its limitations; healthy forests are essential for more than their carbon storage. They are critical for clean water, healthy soil (where a lot of the stored carbon can be found), wildlife habitat, species diversity, and recreation. However, these ecosystem services of the forest are not quantified by carbon alone, nor by a full LCA. Nor are social and economic criteria, which are pillars of a holistic sustainability, present in most LCA evaluations.

Support of truly healthy forests and their surrounding communities means that project teams should only source from responsibly managed forest lands. We urge you as a de-signer or builder to research and compare programs within the region of sourcing, and prefer wood from those demonstrat-ing truly best practice. In general, the more prescriptive programs such as those of the Forest Stewardship Counsel ensure carbon sequestration via smaller clear-cuts, longer rotations, more retention, and larger riparian buffers, but practices vary across regions. Where available, Chain of Custody certificates create a robust link between the forest practices and the wood that ends up on the project.

Second, efficient use of wood materials often entails using glues to create engineered composite wood products, chemical preservatives for durability, or flame retardant chemicals painted on the wood surface to meet fire ratings. Some fabrications have been explored that use no glues, but structurally these are much weaker and therefore not suitable for most applications. However, most structural wood glues have either low or no VOC (Volatile Organic Compound) emissions and therefore should not be a major concern to specifiers. More chemically benign processes also exist to improve durability without the use of toxic chemicals, such as thermal and chemical modification. Yet far too many product ingredients are unknown, making it difficult for specifiers to choose the more benign options. Awareness of these issues throughout the decision process can help avoid using harmful substances while designing for strength, durability, protection of air quality and indoor environment, and protecting those affected upstream and downstream in the product life cycles.

Notwithstanding all that, LCA can still be a useful tool, as it was for the two studies Arup performed for the US Tall Wood demonstration projects. Here are a few other caveats when using LCA or looking at other LCA studies:

- The functional unit is not always clearly defined, or is defined by mass instead of by the actual function such as the floor area, or entire building system, meeting a certain design criteria. Unclear or inappropriate definitions by wood proponents can lead to weak claims and allow competing material industries to criticize environmental claims of wood solutions.
- Uncertainty and variability are often not published within LCA case studies of buildings, and taking them into account when interpreting LCA results often requires some expertise. LCAs must take the burden off designers and make discussion of uncertainty and variability a norm. (See also Chapter Two: Counting Carbon).
- At the product level, request Environmental Product Declarations (EPDs); these aid hugely in understanding impacts and selecting preferable products. More importantly, requiring EPDs fosters greater transparency and the use of LCA in the buildings product industry as a whole.
- Even "all-wood" projects inevitably use some concrete and steel (rebar) in their foundations. In keeping with low-carbon ambitions, concrete should be specified to very low levels of cement (see Chapter Six: Concrete). One should favor ready-mix suppliers who have "off-the-shelf" low-carbon mixes, and employ sustainable plant practices and procurement up the supply chain.
- Design teams can and should account for the various factors that influence the

long-term carbon storage capacity of a project's wood products. This includes differentiating between product types and their application, species selection, and credibility of certifications. It also means considering activities at the different life cycle phases of the wood material, such as aspects influencing service life and the fate of the material at end of life. At the very least, specifications should require that mass timber components be permanently labeled with their material properties and/or governing specifications, facilitating future reuse.

The Future

We expect that completely timber buildings (rather than hybrids) are likely to be most cost competitive in the 6–12-story range. At this height, wind comfort, fire resistance, and structural strength can be achieved with relatively slender CLT panels and simple connection details. Above about 12 stories, wind comfort, progressive collapse resistance, and consequently the structural connections become significantly more complex, so such structures will generally have steel or concrete stability cores. (See also Chapter Ten: Size Matters.)

In this range, the use of timber (rather than carbon-intense steel or concrete) can act as effective carbon storage. According to a study by Dovetail Partners, based on the square-footage of buildings constructed between 4 and 10 stories in the US in 2012, and on projected growth rates in 2015, the near-term potential increase anticipated in using wood for buildings above 6 stories is approximately 1.6–2.4 billion board feet. This equates to 7–10 million metric tons of annual carbon benefit, which is like taking two million cars off the road, or shutting down two or three coal-fired power plants.[1]

Timber buildings can and should develop a larger portion of the total building stock than currently, particularly low and mid-rise. Doing so could create a huge reduction in the carbon impact of the construction industry, while still remaining within the sustainable capacity of the world's managed forests.

With the race to design taller wood buildings, robust strategies to resist fire, progressive collapse, and (where relevant) earthquakes, as well as controlling lateral vibrations, are absolutely vital. Just as these issues were resolved in the past for tall steel buildings, so now current research is helping to resolve them for timber. It's an incredibly exciting time that will see taller timber structures become a permanent feature of our urban landscapes, offering the industry a paradigm shift in how carbon can be an input, not only an output, of what we build. ∎

Seeing the Forests for the Mass Timber

by Jason Grant

Forests are the lungs of the world, breathing in carbon dioxide and exhaling oxygen. Trees use sunlight to convert CO_2 to sugar, which is carbon based. Some of the sugar is used immediately for energy to fuel tree growth, converted back to CO_2, and released back into the atmosphere. The rest combines with other ingredients to make wood, whose chemical composition is about half carbon that will remain safely socked away until the tree dies or is cut down, and its wood rots or burns.

Trees, then, are solar-powered carbon sponges, and although the carbon they soak up eventually cycles back into the atmosphere, the emissions associated with wood are far less than with most other building materials whose extraction and processing sequester little or no carbon, and whose manufacture entails burning lots of fossil fuels. Given the levels of greenhouse gas emissions (GHG) associated with the production of metals and Portland cement, it's no wonder that few look beyond the assertion that we can and should fight climate change by replacing steel and concrete with wood in building structures wherever possible.

But it's not so simple. The case for mass timber buildings should address more than wood's intrinsic virtues; it becomes even more compelling when the timber is ecologically grown and harvested. Forest management, restoration, and conservation are huge parts of the equation, and no honest accounting of the embodied carbon of wood products can ignore them.

Otherwise stated, when it comes to carbon, not all wood is equal: wood may be good, but wood from good forestry is much better. This is true not only when it comes to combating global warming but for myriad other reasons as well.

In our search for solutions to the climate crisis, we tend to miss the forest for the timber. If our main goal is to sequester carbon, one of the best things we can do with forests is to keep them around and let them grow old. Older forests generally capture and store far more carbon over time than do younger ones. And the oldest forests store the most: what remains of the Pacific coast's original ancient forests are among the greatest carbon sinks on the planet.

A team of researchers from Humboldt State and the University of Washington spent seven years painstakingly measuring the carbon stored in the old-growth redwood forests of Northern California, counting trees and understory plants, trunks, branches, needles, and leaves in a dozen plots, then extrapolating to arrive at the most accurate estimates of forest carbon to date.[2] They found that these stands of enormous trees contain 2,600 metric tons of carbon per hectare — two and a half times as much as the old-growth fir and cedar forests further north and about ten times as much as is typical in tropical forests! The imperative to stop stoking global warming by keeping forest carbon captive is thus one more good reason to preserve what virgin forests are still around.

There is, however, plenty of second- and third-growth forest out there and available for harvest; that is where to find the wood for mass timber architecture.

The big question, then, is not whether but *how* working forests will be managed. The trend toward mass wood construction has been fueled by a forest products industry eager to grow market share at the expense of other materials. But most of North America's largest timber companies are wedded to a model of intensive industrial forestry that is far from climate friendly — a fact ignored or at best glossed over by most of mass timber's proponents.

The term "industrial forestry" is environmentalist shorthand for a set of techniques aimed at producing as much commercially valuable timber per acre as quickly as possible, including large-scale clearcutting, replanting trees in monocultures, and heavy applications of herbicides and fertilizers. The amount of time between timber harvests is known in forestry terms as a rotation, and rotation length can vary significantly depending on tree species, geography, and the approach taken to forest management. Generally speaking, industrial forestry favors shorter rotations — say 30 to 50 years in temperate regions — over longer ones.

From a short-term financial perspective, short-rotation industrial forestry is understandable. Forest managers have powerful incentives to harvest timber as soon as it is merchantable. This is mainly because of opportunity cost: when companies sink capital into timber, they cannot invest it elsewhere. Money invested in stocks and bonds will generate income on a regular basis; investment in trees has to wait decades to generate a return. Capital is tied up for so long that it becomes a deterrent to allowing trees to grow older. Optimal capital management and optimal carbon management are conflicting goals for so long as carbon emissions are untaxed. (See "A Price on Carbon" in Chapter Eleven: Action Plan.)

And yet long rotations and lower-impact logging is exactly what is needed if we are serious about tackling global warming. The preponderance of evidence indicates that forests managed on longer rotations sequester significantly more carbon than do those managed on shorter ones. Several studies have found that in the Pacific Northwest 100-year rotations resulted in twice as much total carbon storage as 30-year rotations. Furthermore, this increase was the *combined* result of carbon stored in forest ecosystems and wood products. This is because, over the long haul, forests that are managed less intensively than normal — using longer rotations, smaller clearcuts, and/or selective harvesting — can produce more wood *and* sequester more carbon in plants and soils.

At the same time, conventional industrial forestry doesn't just store less carbon; it can be a huge source of GHG emissions. This point was driven home by a 2015 study[3] that found that the aggressive logging of Oregon's forests since the turn of the century represented between 16 percent and 32 percent of overall emissions in the state — an amount equivalent to two to four million new cars on the road. The main reasons cited for this enormous carbon release were reduction of forest cover resulting from rapid and extensive clearcutting of carbon-dense older forests.

Much is made of the fact that wood stores carbon for the length of its life in a building, ignoring the fact the carbon embodied in wood products accounts for only a fraction of the overall carbon stored in the forest they come from — as little as 18 percent by one estimate. Of the remainder, large amounts may be released to the atmosphere when logging slash rots and soils[4] are exposed by logging. For decades after they occur, clearcuts emit more carbon than regeneration absorbs in spite of the rapid growth rate of young trees. This is because decomposer microbes in the soil work more quickly after a stand is logged, releasing CO_2 as they decompose branches, roots, and other organic matter. The larger the openings, the faster and greater the release. When it comes to pumping large amounts of carbon into the atmosphere, large-scale clearcutting of older forests may be second only to total deforestation (e.g., clearing forests for agriculture) — more even than catastrophic wildfire.[5]

Life cycle assessment (LCA) is the standard method used to measure and compare the carbon footprint of different building materials (see Chapter Two: Counting Carbon). Unfortunately, LCA is a young, imperfect, and evolving science; most LCA systems in use do not account for the variation in the amount of carbon capture, storage, and emissions associated with different approaches to forest management. Through the lens of conventional LCA, all wood looks the same regardless of what happened to the forest of origin. While in general wood bests steel and concrete when it comes to embodied carbon — a ton of carbon in a tree is 3.67 tons of CO_2 removed from the air (see opening section "A Word about 'Carbon'" in the Introduction) — in some cases the magnitude of that advantage would be much less if LCA were more comprehensive. Just as bad, conventional LCA does not currently address harm that logging may cause to the integrity and diversity of forest ecosystems, to water quality, or to threatened and endangered species.

Most experts contend that LCA's omissions are not due to any deliberate attempt to deceive, but rather because LCA is the wrong tool for the job — it doesn't have the capacity to reliably measure forest carbon or ecological impacts, much less regulate them. Many point to a different tool for addressing not just the environmental but also the social impacts of forest management: *forest certification*.

Inspired by the success of organic agricultural certification, the concept of certifying forestry and forest products took root in the late 1980s and early 1990s and gave rise to the founding of the first global forest certification system, the Forest Stewardship Council (FSC). FSC established standards for responsible forestry — a set of requirements that include sustainable timber harvest levels, preservation of wildlife habitat and old growth, protection of rare species and ecosystem types, reduction of herbicide use, and respect for the rights of indigenous peoples as well as the economic well-being of forest-dependent communities. Then FSC created the systems by which forestry operations are independently and annually audited against these standards. Those that meet and adhere to them earn the right to market their products under FSC's green seal of approval.

Though they have overarching consistency, FSC forest management standards vary by region and forest type at the detail level. In general, though, FSC requires a lighter touch than is typical of industrial forestry operations that operate at or near the regulatory floor: smaller clearcuts (where clearcutting is the management technique of choice), more widely distributed across the forest landscape; broader buffers along waterways, and hence more retention of living trees; and, in the western US and British Columbia, longer rotations. In other words, for reasons already explained, FSC requires and rewards more carbon-friendly approaches to forest management.

That less intensive, more selective logging keeps more carbon out of the atmosphere is also a fact recognized by emergent carbon markets. In California, for example, such a market was created through groundbreaking climate legislation, the California Global Warming Solutions Act of 2006. This law mandates a reduction of California's greenhouse gas emissions to 1990 levels by 2020. Among other things, the law created a cap-and-trade program that requires utilities, large industrial plants, and fuel distributors to buy carbon credits from registered carbon projects. In the forest sector, these projects can garner credits through urban forestry, reforesting previously cleared lands, and through improved management techniques that mirror many of those

required by FSC. It's no accident that many of the registered forest carbon projects in the state are FSC certified.

However, FSC isn't the only forest certification system around. Shortly after FSC's founding, the major timber companies launched a competing system through their trade association, the American Forest and Paper Association, called the Sustainable Forestry Initiative (SFI). Two decades later, SFI certifies more than one and a half times as much forestland in North America as FSC, including the holdings of the largest timberland owners on the continent. SFI's toughest critics are environmental groups who accuse it of greenwashing status quo industrial logging at or near the regulatory floor. Its supporters, on the other hand, point to improvements that SFI has introduced to the industry writ large, and argue that any forest certification is better than no certification at all. Still, insofar as FSC holds a higher bar when it comes to clearing sizes, rotation length, and levels of retention, there is no question that FSC is the superior system when it comes to carbon and forest ecology.

* * *

The prospect and promise of mass timber buildings becoming widespread is exciting. The possibility that "tall wood" could be connected to excellent forestry — that the drive to shrink the carbon footprint of buildings and cities could underwrite the recovery of healthy ecosystems, the return of salmon runs, the gradual extension of old-growth, and the sustainable livelihoods of generations of forest industry workers — *this* is downright inspiring.

Fig. 4.8: Not all "sustainable forestry" is equal. This photo is of SFI-certified lands where conventional, industrial forestry is practiced surrounding a still-lush FSC-certified forest near the Pacific coast of North America. Large-scale clearcutting and replanting on short rotations reduces biological and ecological diversity and diminishes the carbon-storing potential of the forest and forest soils. In this rainy climate and mountainous terrain, it also contributes to soil erosion, stream sedimentation and increased risk of landslides. SFI certification may be better than none, but it is still much less protective of the forest and carbon than FSC. *(See also photos in colour section.).* Credit: Photo: Google Earth Delineation: Jason Grant ∎

Notes

1. J. Bowyer, et al. 2016. "Modern Tall Wood Buildings: Opportunities for Innovation." Dovetail Partners, Inc.

2. R. Van Pelt, et al. 2016. "Emergent crowns and light-use complementarity lead to global maximum biomass and leaf area in *Sequoia sempervirens* forests." *Forest Ecology and Management*, 375, 279–308.

3. Talberth, J., D. DellaSala, and E. Fernandez. 2015. *Clearcutting our Carbon Accounts*. Global Forest Watch Report, Center for Sustainable Economy.

4. You may not think of the ground beneath us as a potential source of GHG emissions, but it can be. As they age, forests accumulate increasing amounts of carbon in the soil that supports them; scientists estimate that well over half of the carbon stored in US forests is located in the soil. When a forest is logged, some of this carbon is inevitably released to the atmosphere.

5. A number of studies have shown that even the most destructive fires release less than a third of forest carbon, and some release as little as five percent. Most of the carbon is transferred from living to dead pools but remains onsite — as burned tree boles, for example — and is then taken back up into plants as the burned area regenerates.

Chapter Five

Straw and Other Fibers: A Second Harvest

with Chris Magwood and Massey Burke

W<small>E ALL LOVE FORESTS</small>, those lovely and powerful symbols of climate health. Their ability to soak up atmospheric carbon dioxide and convert it into wood makes them the poster children of carbon sinks, a fact much promulgated by the wood products industry. (See also Chapter Four: Wood.)

However, there are other carbon-capturing "forests" that have so far escaped much attention — the vast miniature forests where we grow our food. Over 720 million hectares of cereal grains were grown worldwide in 2014,[1] an area about the size of Australia. We grow these plants, such as wheat, rice, oats, and barley, to eat their seeds. But supporting

every nutritious seed head is a tubular stalk that is essentially a little tree. Just like their larger cousins in the forests, these little trees — commonly called straw — also contain a lot of carbon. The amount of carbon in straw varies based on the type of plant and the growing conditions, but anywhere from 35–60 percent of straw's mass is carbon. And because it is a lot quicker and easier to grow a field of grain than a forest — we're doing it all the time no matter what — straw-producing agricultural ecosystems are a CO_2 bioabsorption system that we already have functioning around the world.

When you hold a single piece of straw, it seems unlikely that this miniature tree trunk could have any effect on the world's carbon balance. But some rough math — three tons of straw per hectare at 40 percent carbon by weight — gets us to almost a gigaton (Gt) of carbon (or 3.7 Gt of CO_2) being removed from the atmosphere and stored in straw every year.

Further, this total doesn't include the carbon content of other crops such as oilseed and legumes, which also produce vast tonnages of usable fibers. Nature is constantly pulling CO_2 out of the air with the very plants we grow to eat.

To put that 3.7 Gt of CO_2 into perspective, it is equivalent to about 10 percent of the world's CO_2 emissions from fossil fuels (35.9 Gt in 2014[2]) and more than the 1.9 Gt emitted annually by the transportation sector.[3] That's a lot of carbon! A lot of this straw is reincorporated into the agricultural soils, but a lot of it returns to the air as CO_2, either by burning or decomposition.

For a number of reasons, agriculture is itself a major carbon emitter, but if we can retain some of that plant carbon in the straw, we will have created another in an array of carbon sinks to ameliorate climate change. What is needed, then, is a place to put the straw for a very long time to sequester all that carbon. What better place to look than the industry that uses far, far more physical stuff than any other: construction.

Straw has three qualities that make it useful to builders: it is lightweight, the stems are hollow and strong, and it is abundant — so long as we farm and eat, we'll have lots of straw. Through most of history, straw was mostly just left in the fields or burnt to decompose and fertilize the next planting. Its use in buildings was limited to fibrous reinforcing for adobe bricks, thatching for roofs, or for stuffing into timber walls as insulation. With the dawn of industrialized farming, however, that began to change; fields were larger, crops often rotated faster, and burning became a pollution problem for nearby towns and cities. We had to move straw around, and responded with the invention of horse-powered baling machines in the late 1800s. It was then a short and easy conceptual leap to imagine stacking the bales as walls, and European settlers in the Sand Hills of Nebraska did just that in a region with no stone, wood, or even usable sod. Straw bale construction was born, and over the following century has spread and evolved in fits and starts all over the world.

In the past few decades, paralleling the advent of the straw bale revival, inventors have offered us a large and growing array of straw building products more tailored to the standardization needs of the building industry, and to the needs of urban dwellers with limited space. Those products mostly fall into one of several categories:

1) Straw bales and bale panels: prefabricated, typically within wood frames
2) Straw blocks: stackable, for insulation and/or structure
3) Straw panels: for insulation and/or structure
4) Plant fiber insulation systems, sprayed or packed, often using binders such as clay or lime

All of these systems do either or both of two things. One, they can mimic wood by binding the straw fibers together, simulating the cellulose/hemicellulose/lignin structure of wood. When the fibers are closely packed and oriented, and the glue is strong, you get something very much like wood in its mechanical properties. When the binding is loose, as in a straw bale, the strength is less yet still able to carry surprising load. Two, they mass the straw in a way that takes advantage of its tubular structure so as to insulate.

Most insulation systems depend on trapping air — the real insulator — in small enough chambers so as to prevent convective loops. That's how a down jacket or fiberglass batt or spray polyisocyanurate foam all work. That's also how birds' feathers and mammals' fur keep their owners warm: it's all about skillfully trapping air. Straw does that too, and we're only just starting to study straw as a building material, and create new products that use it.

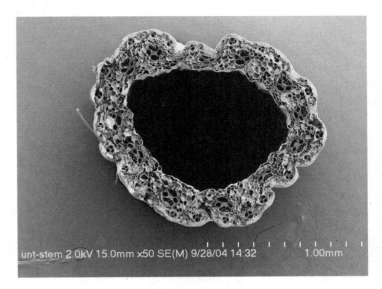

Fig. 5.2: *Straw under a microscope; it's strong and it insulates because it is tubular.* Credit: Courtesy Delilah Wood, USDA-ARS-WRRC

Here's a cursory review of a fast-evolving straw building product industry.

Straw Bales and Straw Bale Panels

Bound rectangular bundles of straw first appeared in the late 1800s — the first straw building products, though not intended as such — and many of those first straw bale buildings in the Sandhill region of Nebraska are still in use. The technique of building with straw bales was rediscovered in the early 1990s, and in the following decades has moved from a fringe alternative to receiving a chapter in the International Residential Code. Thousands of straw bale buildings now exist around the world, in at least 50 countries spanning all the settled continents.

The engaging simplicity of stacking up large rectangular blocks of affordable insulation material continues to attract many

Fig. 5.3: *Keeping a straw bale wall plumb and straight before plastering.* Credit: Chris Magwood

homeowners to straw bale construction, and construction companies familiar with this technique continue to make inroads in the (mostly) residential market. Straw bale construction has made its way from mainly rural locations to urban and multifamily structures, and is a mainstay of a handful of affordable housing organizations.

Despite its many attractive qualities, including a very high carbon sequestration to emissions ratio, site-built straw bale homes are unlikely to become a mainstream construction norm. Working with straw bales

requires builders to rethink their supply chains, retrain workers, and redesign the basic framing system used in conventional construction. While proven to be as cost effective as conventional construction, this kind of shift in the industry seems unlikely to happen on a large scale. But for the homeowners and builders who are committed to building with the most carbon sequestration and the lowest operational carbon footprint, this continues to be an attractive option.

Prefabricated Straw Bale Wall Panels

The labor and supply chain issues noted above can be largely addressed by the prefabrication of straw bale wall panels. There are already several companies around the world doing this on a commercial scale. Probably the most advanced is Ecococon, producing computer-designed and fabricated panels at a factory in Lithuania and supplying projects all over Europe. Others are operating in Australia, Canada, and England, but each uses the same basic system: packing bales tightly within sturdy timber frames for transport and assembly onsite. In this form, straw walls become practical to use for any developer or contractor, as there is no need to retrain workers to deal with straw bales. Any construction worker can install a prefabricated straw bale panel, and the costs, even at the current small scale of production, are competitive with and sometimes better than conventional construction.

Whether site built or prefabricated, straw bale walls provide high thermal performance for multiple reasons. Mainly, they are thick; the thermal resistance per inch is only about

R1.4–2.0 per inch, roughly half that of cotton batts or blown cellulose, but the bale itself is big. Depending on what kind of bale it is, and how it is stacked, the wall assembly will be anywhere from 14 to 24 inches thick, making for a total package greatly exceeding any code-stipulated thermal insulation requirements. Also, because the necessary plastering inside and out can make for a much more airtight assembly than wood-framed walls, bale walls don't leak and as such are popular for Passivhaus and net-zero projects.

Straw Blocks

Sometimes called Straw Bale 2.0, the idea with blocks is to compress and bind the straw with adhesive into shapes that are more consistent, and more dimensionally compatible with other elements in common structures like wood-framed houses. That's the idea, anyway. Several firms promoting one version or another have come and gone, and none are available on the market as yet. Products we have seen have good load-bearing capacity, and thermal insulation values of about R2 per inch.

Hempcrete straw blocks, using hydraulic lime as the adhesive, are beginning to enter the European market, and one Canadian start-up is beginning to prototype this kind of straw block in North America.

Straw blocks are relatively simple to create, and the modular form is easily adaptable with existing building trades. It is likely that straw block will continue to attract inventors and developers until some or several forms of block succeed in the marketplace.

Straw Panels

There are two basic types of straw panels. The first, based on a decades-old Swedish technology known as Stramit, is an insulating, lightweight panel made by compressing straw with heat so as to release innate lignin binders. The resulting panels are bound in kraft paper and produced in thicknesses from two to six inches, weighing around 22.5 pounds per cubic foot. Their thermal

Fig. 5.4: *Compressed stackable straw blocks.*

Fig. 5.5: *Next generation oriented strand board — made of straw.*

resistance is modest, ranging around R1.4 per inch, and their compressive strengths are minor. They are currently manufactured in North America under the trade name Durra.

The second type of panel is very much like the well-known wood Oriented Strand Board (OSB) commonly used around the world. The technology for making them would seem to be much like that for making OSB, but because straw is tubular — making it difficult to fully glue — the panels never reached their promise until Canadian-sponsored research led to a way to split the straws before gathering, orienting, and adhering them into panels. Many versions have now appeared around the world, and

Fig. 5.6: *Insulating thatched straw walls of the net-zero Enterprise Centre, University of East Anglia, Norwich, UK.* Credit: Courtesy University of Bath

have properties comparable to or better than OSB for strength and durability.

Bonded Plant Fiber Insulation Systems

Straw and other plant fibers have historically been used for insulation, such as the straw-clay infill for Germany's famous medieval timber frame structures, or the thatched roofs of England, Africa, Japan, and many other places. Those techniques are being kept alive and even promulgated in various parts of the world, but really only make sense where labor is cheap. The promise of straw insulation has prompted much innovation all over the world, and a dizzying array of products has been growing for several decades.

Among these many products, two stand out: an old favorite, and a brand new version of the same idea. The old favorite is straw-clay, being revived in many places as hand-packed or machine-sprayed slurry; it is a simple mix of chopped straw and a syrupy clay slip that dries in place after taking the desired shape. The ingredients are cheap and plentiful, and the clay serves to protect the straw from insects, fire, and decay. It's a lovely and elegant technology using materials available almost anywhere, and can become even better and more effective with more research and testing.

The second product builds on the same idea, but with materials specifically selected for the purpose: hemp hurds (the part of the amazing hemp plant left over after other users have taken the finer oils, seeds, and fibers), and a hydraulic lime binder. Like straw-clay, the hurds are dried, chopped, and mixed with the lime, then hand packed or sprayed into wall and roof cavities. The resulting matrix is stronger, more thermally effective and durable than straw-clay, and is coming into popularity in the UK, France, and Canada. It appears that as much as 50 percent of the carbon emitted during firing the lime is, within a year or two, reabsorbed as the lime cures (a process called carbonation, that by definition absorbs CO_2), so this is very much a climate-friendly technology.

Clearly, significant reductions in carbon footprint can be achieved by replacing high-carbon materials like foam and mineral wool with sequestering materials like wood, hemp, and straw. These materials also *reward* the use of more insulation — adding more thermal performance using sequestering materials further reduces the embodied carbon footprint of the building and reduces the operational carbon footprint.

Plant fiber products in general have distinct advantages and disadvantages compared to their competitors. The big advantage is that they are cheap and plentiful, and sequester carbon that would otherwise end up in the air. The primary disadvantage is their susceptibility to moisture decay. Like wood and wood products, straw fiber products are cellulosic and can provide food for mold and fungi. However, various control strategies at the material, assembly, and building scales have been worked out as the industry has gained experience with wood products to manage that risk. In this regard, plant fiber building products present the same moisture risks and design issues as wood products such as particleboard and plywood.

Fig. 5.7: *Hempcrete, also called Hemcrete or Hemp-Lime, can be hand packed, sprayed, or formed into blocks, providing both insulation and thermal mass.* Credit: Chris Magwood

Fig. 5.8: *Bamboo can be milled and fabricated into very high-strength I-beams, panels, and other shapes.*

It's something to pay attention to, but not difficult to manage.

This is an inevitably incomplete list, not just of ideas, firms, and specific products, but of developments in the use of agricultural by-products in construction. The possibilities

Carbon Emissions and Sequestration[4]

The embodied carbon of a building material is the total of all greenhouse gas emissions resulting from its harvesting, transportation, and manufacturing processes. All building materials and products have some amount of embodied carbon, and this figure is expressed as $kgCO_2e/kg$ — kilograms of carbon equivalent per kilogram of material. Factors for $kgCO_2e$ are found in all embodied carbon databases.

What is much less likely to be found in a database is the carbon *sequestration* of a material. Plant-based materials "digest" atmospheric carbon during pho-tosynthesis such that the carbon constitutes 35–60 percent of their dry weight, depending on the plant type and growing conditions. Left in the biosphere, this carbon content is released again as the plant decomposes, but if it is used in construction, it is removed from the atmosphere and is sequestered in the building. For materials like wood, straw, hemp, bamboo, and cork, this can result in vastly more carbon being sequestered than was emitted for harvesting and manufacturing, as seen in the table below:

Insulation material	Embodied carbon (by weight)[5] $kgCO_2e/kg$	Embodied carbon for 4x8-foot wall @ R-28 $kgCO_2e$	CO2e sequestered (-) or emitted (+) for 4x8-foot wall @ R-28, kg per panel
Hempcrete	0.142	45	-87
Straw bales	0.063	22	-78
Cork	0.19	13	-46
Dense pack cellulose	0.63	41	-19
Denim batt	1.5	22	+2.6
Fiberglass batt	1.35	18	+18
Mineral wool batt	1.28	22	+22
Expanded polystyrene foam	3.29	37	+37
Extruded polystyrene foam	3.42	39	+39
	*Figures from Inventory of Carbon and Energy V."	**Material densities from *Making Better Buildings*	

and prototypes are many, but the infamous "valley of death" that any start-up company must traverse is particularly long and deep in the world of building materials. Why? Because it takes a very great deal of both capital and time to develop a product, pass necessary standards and code acceptance, and market it successfully in a famously conservative, hostile, and risk-averse industry. Time and again we have seen venture capitalists take a look and pass because other investment opportunities can pay out in 2 to 5 years, much more appealing than the 10 to 20 years needed in construction. There is plenty of opportunity for research and innovation with the many agricultural by-products being produced every year, but it may take a meaningful price on carbon in the marketplace (see Chapter Eleven) to invite the necessary and patient capital that can foster successful developments.

In this conversation about moving carbon in the right direction — that is, out of the atmosphere and into buildings — it's worth looking at the supply chain for straw: modern industrial agriculture. To date, most case studies examining the carbon sequestration capacity of straw consider it a by-product of food production. In this scenario, the only emissions assigned to straw production come from baling and transport. Since describing the boundary conditions for carbon footprinting any material is notoriously difficult, this is not a bad place to start. However, a closer look at the emissions associated with straw is important both for intellectual honesty and for more accurate carbon accounting. Straw is a waste product in modern agriculture because modern agriculture is, from an ecological perspective, a rather bare-bones, linear approach to a food-growing landscape. It needs a lot of carbon-emitting inputs to produce a very narrow range of outputs. Though beyond the scope of this book, it bears at least noting that many climate-friendly innovations are appearing in agriculture: elegant, complex grain-growing systems that yield multiple outputs as well as multiple opportunities to sequester carbon. These systems are much simpler than the grassland ecosystems that inspire them, but are much more complex than the industrial ones that produce most of our grain. A well-designed polyculture grain grassland will

sequester and build soil carbon over time; and when the grain is harvested, the whole system yields straw to be sequestered in the built environment. As the polyculture becomes older and more established, additional carbon sequestration loops can be built in. Even beyond this, the real power of designed grassland-based carbon sequestration strategies is that they develop quickly and yield information quickly. A carbon-sequestering, food-producing grassland can be designed and implemented in one year, and then adjusted or altered the following year, as needed, to increase yield and sequestration, and respond to changing climatic conditions. By contrast, a carbon-sequestering forest takes many years to establish. Neither is better than the other; but we need as many tools as we can find or invent to rebalance the global carbon cycle. Grasslands and forests are complementary carbon sequestration systems, just as straw and wood are complementary materials within the fabric of a building. A healthy long-term approach to both, economically and ecologically, is to study and foster the soils from which they spring.

Climate change challenges us to reorient the way we work with ecological cycles, globally, locally, and at every scale in-between. To respond effectively to climate change, we have to learn to think like an ecosystem, and to design what we make accordingly. Straw is a humble and unassuming material, yet it is also one of the most direct links between the human economy and the global carbon cycle; we are only just learning how to use it creatively. Most of the excitement is still to come. Stay tuned.

The Planet's Sixth Carbon Sink: A Success Story

by Craig White

We can decarbonize the physical design of our buildings by seeking out those materials that are less energy and carbon intensive. This can be done by ensuring we use, for example, recycled rather than original metal work, specify Portland cement replacements and recycled aggregates in our concrete, and seek out materials that require less processing and transport. At first, quick wins can be secured, but we quickly see diminishing returns on these initial and easy to make specification changes. We then arrive at a decarbonization plateau where, without significant innovation, further reductions seem beyond reach. There is, however, a new and emerging opportunity to switch to a new materials palette for use in mainstream construction. We can switch to a renewable materials palette by building with carbon.

We should first define what we mean by non-renewable materials. Many materials used in construction are obtained from finite resources that, once exploited, can never be replaced. They are not renewable — at best they can only be recycled in perpetuity. While the planet has an abundance of non-renewable resources, extraction costs are rising, the environmental impact of securing them is increasingly unsustainable, and their availability is decreasing.

Renewable materials are generally accepted to mean those materials that are biogenic in origin. For example, a historically understood and used renewable material is timber, but there are many others. They include: cotton, flax, hemp and other vegetable fibers, cork, wool, straw, thatch, and bamboo. There are less well-known ones such as: agar, sea grass, silk, beeswax, lacquer, linseed oil, shellac, tar, turpentine, rubber, casein, resin, starch, rice, and peanut hulls. What makes these materials renewable is that they are sourced from living plants and animals — not from mineral deposits or fossil deposits of organic material such as peat, oil, gas, and coal.

Biogenic renewable materials are solar powered, or, more specifically, created through photosynthesis. Plants are able to grow through the conversion of atmospheric CO_2 into usable carbon. Powered by photosynthesis, plants absorb atmospheric CO_2, deconstruct the molecule into its component atoms, carbon and oxygen, and bank the carbon into complex sugars, the building blocks of cellulose. This natural Carbon Capture and Storage (CCS) process allows us to exploit the physical properties of renewable materials, such as timber and straw, to build with carbon. Instead of simply reducing the embodied carbon of materials we use, we can switch to renewable, carbon-banking materials, and through scaling their use, we can not only build with carbon, we can think of our built environment as a carbon sink, the planet's sixth carbon sink.

Having started this journey 15 years ago, our company has developed a system of construction that allows for the prefabrication of straw and timber construction panels, called ModCell (www.modcell.com). The amount of CO_2 stored in the ModCell system is more than is emitted through its making and transportation, resulting in a carbon "banking" footprint. This CCS mechanism means we are able to deliver a scalable system of renewable, carbon-banking construction. The ModCell system has been used on over 90 projects across the UK in schools, offices, community projects, and housing.

The LILAC Cohousing project in Leeds is an example of scaling the use of renewable materials in the housing sector. LILAC is the UK's first affordable, low-carbon cohousing project, set up to demonstrate how community-led development can deliver sustainable living. LILAC means Low Impact Living Affordable Community. It is a member-led, not-for-profit cooperative society and has built a community of 20 ModCell homes in urban Bramley, on the site of an old school. The LILAC walls, roofs, floors, and partitions are built entirely from timber, straw, and lime render. (Lime render produces much less carbon in its manufacture than cement render.)

Just how much CO_2 is banked in the renewable materials used at LILAC?

Timber typically captures 742 kilograms of atmospheric CO_2 per cubic meter of timber.

Straw captures 1.62 kilogram of atmospheric CO_2 per kilogram of straw.

Comparison of emitted CO₂ and captured CO₂ using renewable materials

Typical English masonry house

Light wood frame house

BaleHaus at Bath

BaleHaus with CSB (compressed straw board)

■ Embodied   ■ Timber CO₂e   ▨ Straw bale CO₂e   ▨ CSB CO₂e

Fig. 5.9: *The amount of atmospheric CO₂ emitted and captured by different forms of construction. The gross amount of CO₂ captured is shown as a negative figure.* Credit: ModCell

When two of these materials are combined into a typical 3-meter-by-3-meter ModCell panel, the total amount of CO_2 banked per panel is 1,400 kilograms. When all the captured carbon stored in panels used in the walls and roofs at LILAC is combined with the timber used in the floors and partitions, the LILAC development captured and stores over 1,080 tonnes of atmospheric equivalent CO_2.

LILAC is an example of the built environment as a carbon sink. If all new homes in the UK were built using similar methods, we could capture and store over 10 megatons of atmospheric CO_2 year on year. ∎

Fig. 5.9: *The LILAC cohousing project in Leeds.* Credit: ModCell

Notes

1. data.worldbank.org/indicator/AG. LND. CREL.HA

2. The Global Carbon Budget 2015, Version 1.1 (Global_Carbon_Budget_ 2015_v1.1.xlsx)

3. www.ipcc.ch/pdf/assessment-report/ar5/ wg3/ipcc_wg3_ar5_chapter1.pdf

4. Excerpted from Chris Magwood, *Essential Sustainable Home Design*, New Society Publishers, 2017.

5. www.circularecology.com/news/the-ice-embodied-carbon-database-is-now-hosted-for-download-by-circular-ecology

Chapter Six

Concrete: The Reinvention of Artificial Rock

with Fernando Martirena and Paul Jaquin

What Is Concrete?

CONCRETE IS ARTIFICIAL ROCK: you mix up some sand and gravel with some kind of glue, and pour it or spray it or pack it into whatever form you want to get the desired shape when it hardens. That's how you make a concrete block, an adobe brick, a sheet of gypsum board, a city sidewalk, or the Pantheon — to name just a few among thousands of examples. That's how you build most of the buildings that human beings have ever built — with sand and gravel glued together. There might also be some fibers or chemical additives or other interesting ingredients, and it gets much, much more complicated than you might think, but that's the gist of it: gravel and glue.

"Concrete" as most people think of it — and as modern building codes require — is sand and gravel glued together with Portland cement, which is basically made by heating and grinding limestone. When it was invented in England 200 years ago, Portland cement was a technological leap forward from the lime plasters that had been around in many forms all over the world for millennia. The inventor, Joseph Aspdin, found by trial and error that by burning the limestone hotter (over 1400°C) and intermixing ground clay, he could get a vastly better product. It set up faster, was stronger quicker, could be used underwater, and in many ways was just better than anything we'd had before. And so it took over, and is now the most commonly used building material in the world after sand and gravel. It was a truly great and game-changing technology, and is now required by law to be present in anything deemed "concrete." It is embedded in our standards and codes, and therefore intimately part of almost all building projects. So why does it need reinvention? What's the matter with how it is?

The Problem with Concrete

Two very large problems with how it is. One, you put a ton or more of carbon in the air for every ton of Portland cement produced; at that rate, cement production today accounts for about six percent of all anthropogenic global emissions. Two, there's not enough: there's not enough cement-making capacity in the world to take care of the next three billion people due in the next 15 or so years. Even with all the Portland cement and other cementitious materials in the world, we won't be able to place as much concrete as will be needed to house, educate, and shelter everybody.

But wait! It gets worse: we're not just short of cement, we're short of sand and gravel! Most urban areas in the world, all of them expanding, have to import sand and gravel from considerable distance — and at considerable fuel and carbon cost — to supply their concrete, glass, and many other industries. In some places there are even sand *mafias* who will hurt and kill people in order to access prime deposits. Sand mafias!

The Reinvention Is On

You can make concrete with all sorts of glues besides Portland cement: various industrial waste products, certain volcanic soils (like the Romans did), magnesium and other oxides, and all sorts of other things. But very few of them are available in the quantity we need, so of necessity, and even aside from climate considerations, we need to reinvent concrete. We need to rethink artificial rock.

Plenty of people, you'll be glad to know, are working on just that. Their efforts fall into one or more of several categories such as improving/inventing new cement(s) with less carbon footprint (but scalable and not more expensive), inventing carbon-storing

artificial sand and gravel, injecting and storing carbon in ordinary Portland cement concrete (aka OPC), or changing the way concrete products are made.

We couldn't hope to cover all the efforts, worldwide, to reinvent concrete in these ways, so we'll point out a few of the most promising ones. This is a particularly turbulent and evolving field, so we offer apologies in advance for all the hot prospects we will surely omit.

But First, Some Basics

All human modifications to the landscape are made with earth, because everything we build with originated in the Earth's crust, and has its present form as a result of some combination of geological, biological, and/or industrial processes. For example, wood is the result of solar energy and biological growth using nutrients pulled from the soil and carbon from the air (which itself came from the Earth). Structural steel — and all other metals — are the distillation of certain ore-bearing rocks. Portland cement is made by heating and grinding limestone along with various other trace earthen materials. Glass is melted sand. Even oil-based plastics are the product of processing ancient beds of algae that have been transformed into oil by complex geological processes over tens of millions of years. All buildings are *earthen* buildings because every building material we see around us today came, one way or another, from the Earth. Most of the trick with low-carbon building amounts to minimizing the amount of transport and processing you need between harvesting something at

Fig. 6.1: *Multi-story apartments constructed with cob (puddled earth) and adobe, continuously occupied for at least a thousand years.*

its source and placing it into your building. Less fuss, less muss.

OK, great, so how should you make concrete? Let's start with some illuminating history.

The world's oldest surviving buildings date back at least seven thousand years at sites in Europe and the Middle East, and are typically stacked and fitted stonework. Some of those oldest buildings are made with "mud brick," or *adobe* (from the Arabic *al toba*), essentially the first concrete. Still standing today is the famous city of Shibam in Yemen, the "Manhattan of Mud," built entirely of hand-packed earth, continuously occupied for a thousand years, and testament to the surprising durability of clay-based concrete. (Figure 6.1)

Clay-based concrete in its many forms, such as adobe, cob, and rammed earth, survives all over the world as indigenous architecture — and all too often carries the stigma of poverty; who would live in a mud hut if they could afford better?

The answer to that is: lots of people. Builders, architects, and engineers all over the world have begun to rediscover earthen construction (as clay concrete is widely known),

Fig. 6.2: *Unstabilized rammed earth in the Alps.* Credit: University of Bath

both for its visceral aesthetic appeal (when done right) and for its material properties that are in many ways superior to any alternative. You can, for example, moderate humidity and block electromagnetic signals with clay (according to various reports), but not with any other common building material. It is also very cheap, abundant, and very low carbon.

In recent decades we've come to know much more about how clay works, and how best to work with it. Paul Jaquin has some engineering insight about clay, the original cement.

Clay: The First Cement

Paul Jaquin

Clay is one of the oldest building materials, used ever since the first hunter-gatherers settled down in the Fertile Crescent and the Mediterranean. Here we will talk about clay as a building material: how it acts as a binder for aggregates and how it is used in construction. We will talk about air-dried clay, but not about clays which have been heated to change their chemical structure; these are fired or calcined clays and are used for ceramics, fired bricks, and many other things. (See also Fernando Martirena's piece following this on partially calcined clay cement.)

Historical Building Using Clay as a Binder

Clay has been used as a building material ever since the first settlements in the Fertile Crescent of ancient Mesopotamia in about 9000 BC. Those first buildings accompanied the transition from hunting-gathering to agriculture, and were in fact granaries for the storage of food rather than dwellings. In much of historic concrete architecture, clay is the primary binder; from the adobes of Jericho, Ur, and Taos to the rammed earth of the Great Wall of China, the Kasbahs of Morocco, and innumerable monasteries along the ancient Silk Road to the cob (hand-packed earth) of Arabia, Africa, Peru, and England, clay architecture lives on from antiquity.

Clay has been used for thousands of years as a binder, and even though new materials are discovered and used for construction, clay will always have a role as a versatile and viable building material.

What Makes Clay Special?

Clay and *silt* are fine soil particles which form by the physical and chemical weathering of rocks. Technically, all particles smaller than 0.0025 inches are clay-sized, but clays are unique among soil particles because they are electrostatically charged. These electrostatic charges provide additional forces of attraction between the particles and account for the unique properties of clay.

The mineral *feldspar* is a large component of the Earth's crust, and comprises around 60 percent of the exposed minerals in the world. Clay minerals are formed by the chemical and physical weathering of feldspar in the presence of water. Silts, on the other hand, are highly weathered rock particles of any origin, but critically do not contain clay family minerals, and are not electrostatically charged. Silt particles are typically slightly larger than clay particles, and are generally avoided in all types of concrete because they "gum up" the mix. That is, the presence of silts adds enormously to the exposed surface area of all the aggregate particles in a given volume, making the binder (be it clay or cement) have to "stretch" that much further to hold everything together.

Bonding in Clay

For engineering and building purposes, we do not need a detailed understanding of each species of clay mineral so much as a broad understanding of the main characteristics of clay in general. There are lots of different bonding mechanisms at work in clays, and they work at a range of different scales, from the atomic to the (relatively) enormous size of a silt particle. It isn't important to understand the exact nature of each of these binding mechanisms, but a cursory review of each can be useful.

Sheets, Layers, and Assemblages

Starting at the atomic level, we see planes of silicon and aluminum oxides bond with each other to form sheets. These sheets attract each other to form layers, and then the layers come together to form assemblages up to two microns in size.

Sheets

Silicon oxides form tetrahedral-shaped molecules (triangular-based pyramids), with a silicon in the center and four oxygen atoms at the points of the pyramid. Aluminum oxides form octahedral molecules (two square-based pyramids, with the squares together, to form a shape with eight triangular faces), with aluminum at the centre and the six points of the shape being populated by either oxygen atoms or hydroxide (OH) molecules. The aluminum and silicon oxides share oxygen atoms at the internal join of the sheet, through covalent bonding. This joint sharing of the oxygen atoms allows the two different oxides to bond to one another and form sheets. Note that iron and magnesium can be substituted instead of aluminum and the octahedral shape will not change (this is known as isomorphic substitution).

At the silica side of the sheet, there is a plane of oxygen atoms, whereas at the aluminum side, there exists a plane of oxygens and hydroxides. The arrangement of the

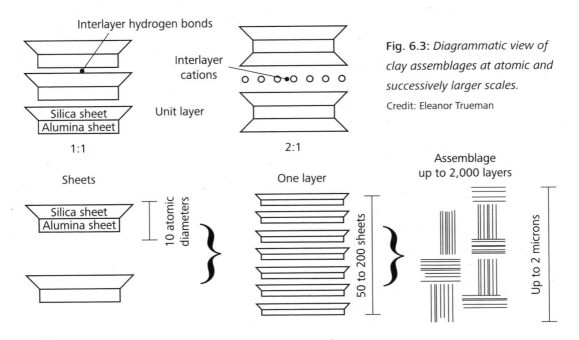

Fig. 6.3: *Diagrammatic view of clay assemblages at atomic and successively larger scales.*
Credit: Eleanor Trueman

covalently bonded sheets determines the type of clay mineral. The three main types of clay mineral are, in ascending order of expansiveness when exposed to water: *Kaolinites*, *illites*, and *montmorillonites*, with each having different arrangements of covalently bonded sheets. In common experience, the stable kaolinites are sought for ceramics and many other industrial uses, while montmorillonites are the scourge of builders because of their extreme volume changes through wetting and drying.

Layers

Bonding between layers occurs via hydrogen bonds from one hydroxide surface to the adjacent oxygen plane of the neighboring layer. Therefore the overall charge of the layer is always close to zero. Van der Waals forces also help to bind the layers together.

The Assemblage

The clay particle assemblage is the scale we are most interested in for engineering and construction. At this scale, a clay assemblage is typically similar in size to a small silt particle (around two micrometers), and other bonding forces come into play which determine the engineering behavior of the material: *friction* and *suction*.

Friction

Friction is relatively simple to understand: when we push an object along a surface, we feel resistance; that force we feel is friction. The size of the frictional force depends on the surface roughness of the two objects and the amount of force which is pushing them

together, such as gravity. The surface roughness depends on the size and shape of the particles, which determine the amount of physical interlock between them. There is also surface-to-surface adhesion, which adds to the friction effect as a result of the electrostatic forces between the clay aggregations.

Suction

The binding strength of clay depends very much on the water in the soil, which holds the clay together as microbridges of water between particles. Unless the soil is immersed in liquid water (such as below the water table, or a clay wall in driving rain or flood conditions), the amount of water present in the soil depends on the relative humidity of the surrounding air.

The liquid bridge exists because of the phenomenon of surface tension. *Surface tension* exists where there is an interface between a liquid and a gas; the surface of the liquid behaves like a sheet in tension and is always curved. You can see this phenomenon in water droplets on a leaf or the surface of liquid in a glass. The curvature of this sheet depends on the relative humidity of the surrounding air.

When we have two soil particles, for example two clay aggregations or two sand particles, then water in the pores between the particles is held in bridges of water, and the water is "captured" by the surface tension membrane. The tension of the membrane and a lower water pressure than the surrounding air act together and provide an attractive force called *suction* which pulls the two particles together. It is most easy to

A closeup look at earth (no organics)

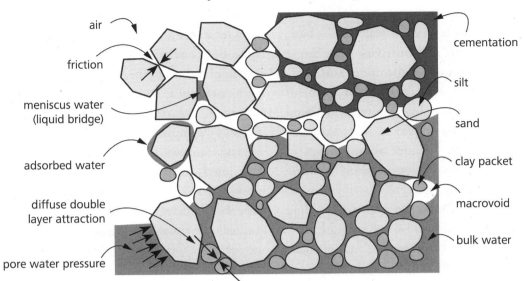

air

friction

meniscus water
(liquid bridge)

adsorbed water

diffuse double
layer attraction

pore water pressure

cementation

silt

sand

clay packet

macrovoid

bulk water

Fig. 6.4: *A closeup look at earth (no organic material).*

see suction when making a sandcastle: with just the right amount of water in the sand, a castle can be built; if the sand is too wet or too dry, it won't hold together. In a sandcastle, the voids between the sand grains are relatively large, so the water soon evaporates from the liquid bridges and the walls come tumbling down. But where clay aggregations are used a binder, the voids between the clay aggregations are smaller and the curvature of the membrane and the relative humidity are in equilibrium, so the water does not evaporate and the castle remains standing indefinitely. Conversely, where too much water is added (for example when the sandcastle or the brick is flooded), then the bridges grow to become continuous water throughout the whole of the material, and there is no longer attractive force between the soil particles. When this happens, the soil loses much of its strength and with enough water will flow as a slurry or mud.

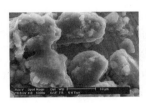

Fig. 6.5: *Water surrounding clay aggregations, with layers of kaolinite visible.* Credit: Sergio Lourenco, University of Hong Kong

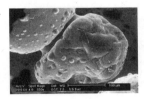

Fig. 6.6: *Water bridge between two sand particles.* Credit: Sergio Lourenco, University of Hong Kong

How Strong Is Clay Concrete?

The compressive strength of a well-built unfired clay brick (adobe) or cob wall is around 300 pounds per square inch, or about a tenth that of a typical Portland cement concrete. Rammed earth, by dint of the mechanical compaction which aids binding, will reach strengths two or three times higher, with comparable increase in its weather durability. Design of clay concrete structures, like any structure, allows for the properties of the material. For example, the buildings in Shibam in Yemen are up to 100 feet tall, and have particularly thick walls at the base, with small rooms and wide walls, but larger rooms and windows on the upper stories.

Much like unreinforced Portland cement concrete, clay is not strong in tension. Good design and detailing, just as with Portland cement concrete, includes adding fibers to reduce shrinkage cracking and erosion, while sometimes adding global reinforcing of steel, rope, mesh, or bamboo to improve building ductility and tensile and shear strength.

In many parts of the world, builders have experimented with adding Portland cement to clay concrete, thus making it, in the parlance of modern building codes, *soil-cement*. The advantages to this are dramatic increases in strength and durability, while the disadvantages are cost, constraints on construction (once you put in the cement, a clock starts running that you cannot argue with), and loss of the humidity-buffering qualities described in the next section. Highway departments the world over mix small portions of Portland cement or lime into clay soils to strengthen and stabilize them against moisture expansion/contraction; from them we now have a fairly rich body of engineering research on soil-cement and soil-lime.

Humidity Buffering and Thermal Mass

One of the major advantages of clay construction is its ability to regulate the relative humidity inside a building by taking in or releasing water vapor (see also Chapter Eight: To Your Health). This happens through the growth (*adsorption*) or shrinkage (*desorption*) of the tiny water bridges between the soil particles which are described above. The ability to buffer relative humidity through the course of a day can be very useful, as when showering and cooking introduce water vapor into the air. We humans thrive most in a relative humidity around 40–60 percent, whereas constant higher humidity can allow the growth of mold spores and dust mites. Using clay as a binder also allows water vapor to be absorbed into the wall and released more gradually, or transported through the wall to the outside. Clay concrete walls also provide significant *thermal mass* — a "thermal flywheel" that evens out daily temperature variations by holding and slowly releasing heat. This, combined with humidity buffering, makes buildings constructed from clay thermally pleasant, and can reduce the requirement for air conditioning.

Future

The use of clay as a binder has seen significantly more research of late as the construction industry grapples with the need to reduce carbon emissions. Many products and processes use Portland cement but don't need to, or not as much; in many cases, clay can provide a complete or partial substitute binder. While clay is often used in "natural" building projects in the industrialized world, such as rammed earth and cob houses, it is becoming increasingly mainstream. Countertops, plasters, and walls in restaurants and showrooms are used for their aesthetic appeal; wall panels can be placed to assist with humidity regulation; and the use of clay binders in structural products is increasing. Prefabrication and offsite manufacture of clay products is also increasing, and the first full-scale 3-D printed clay structures are appearing.

There are of course plenty of things clay cannot do, as is the case for any material, and plenty of structures that couldn't be built well with clay concrete. But plenty can be, and as our knowledge of this ubiquitous, inexpensive resource grows, we hope the building industry and codes will loosen up enough to allow greater use of the ultimate low-carbon concrete.[1] ■

References

www.slideshare.net/smeseguer/the-origin-of-clay-minerals-1-1

Editor's Note

Historically, concrete construction evolved from unfired clay in several directions in many different parts of the world. First, we discovered *fired* clay — ceramics — for making pottery and tiles that could shed rain, endure foot traffic, and just look better. Also, we found that by ramming or compressing the mix in place (see Figure 6.2), we could get much better strength and durability; that is part of the secret of the Pantheon and the Great Wall of China, to name just two. Finally, lime plaster was inadvertently discovered (I believe) when kilns were made of limestone and ended up heating and *calcining*[2] the rock walls of the kiln. Many varieties of plasters and mortars followed, and the Romans found that if they mixed pottery shards into the plaster, they got a stronger, water-resistant concrete. When they mixed their plasters with the volcanic soils of the town of Pozzuoli, near modern-day Naples, the famous Roman concrete was born and eponymous "pozzolans" discovered.

In modern times, engineers have been exploring the range of concrete properties between the relatively weak and soft clay concretes, and the stronger, harder Portland cement concretes so common now in the industrialized world. As of 2016, a very great deal of brainpower and investment capital has gone into discovering or inventing a cement that is as cheap and utilitarian as Portland without the heavy carbon footprint. One promising result comes out of the work of Fernando Martirena in Cuba, using partially calcined clay.

Rethinking Cement

Fernando Martirena

Portland cement is the most widely and massively produced material in human civilization, and by some estimates Portland cement concrete accounts for fully half of global resource consumption. Despite this, cement production and use is associated with only six percent of global greenhouse gas emissions.

Oxygen, aluminum, silicon, calcium, iron, magnesium, potassium, and sodium — cement's main ingredients are the most abundant elements in the Earth's crust. This is why, if you think in terms of scale, cement is the only material that can be massively produced almost anywhere on Earth; the development path for emerging economies goes inevitably through the use of cement as the main building material for infrastructure and shelter.

If humanity hopes to keep global temperature rise below 2°C, carbon emissions in sectors such as energy, transport, and production of materials have to be drastically reduced. As today's measures to decarbonize power generation grow, carbon emissions from the production of cement become ever more prominent, and may expand two to three times by 2050.

In the 21st century, a lot of effort has been made to increase the sustainability of cement. From all the alternatives available, reducing *clinker* — the most important and energy-intensive component of cement — has proven to be the most effective way of reducing environmental impacts. Clinker can be substituted with some materials known as *pozzolans*, usually obtained as waste products from industry; their use reduces the embodied CO_2 of concrete.

Most common clinker substitutes are wastes from industry such as fly ash, a by-product of coal power plants, or ground granulated blast furnace slag, obtained from virgin steel production. But both of these are in diminishing supply. Trends in the energy sector are to phase out coal-based power, so the supply of fly ash will decrease. Virgin steel production (as opposed to recycled steel production; see Chapter Two: Counting Carbon) is concentrated in several industrialized countries such as Japan, the USA, and China, and pressure on energy costs has prompted a reduction in global production; a fall in supply of slag is also expected. The global availability of cement substitute materials is about 10 percent of global cement production, and global clinker substitution in cement has plateaued at around 30 percent in the last few years.

There are, however, other materials which are abundant and whose use could close the gap in the supply of cement substitutes: calcined clays have been extensively used as pozzolans for centuries, from the Romans up to modern times. Large reserves of kaolinitic clays occur throughout the tropical belt of the planet — exactly where the highest growth in demand for cement exists — and the volume of resources available exceeds by far global annual cement demand. Limestone, also a material used in cement production, is another cement substitute material whose reserves would sustain centuries of cement production.

A research team from Universidad Central de las Villas, Cuba, Ecole Polytechnique Federal de Lausanne, Switzerland, and several Indian Institutes of Technology have worked since 2005 in the development of a new ternary cement, called Limestone-Calcined Clay Portland cement (LC3). LC3 is composed, roughly, of clinker and gypsum (50 percent), calcined clay (30 percent), and limestone (20 percent). The LC3 project has been entirely funded by the Swiss Agency for Collaboration and Development, SDC. LC3 cement achieves mechanical properties similar to and more durable than Portland cement.

The idea has moved quickly from the lab to industrial scale. In 2013 the Cuban cement factory Siguaney made an industrial trial to produce 130 tonnes of LC3, and the cement was successfully utilized in the production of concrete and masonry. Further trials have been carried out in India, where the successful performance of LC3 has been validated in a range of applications.

Life Cycle Assessment studies carried out in Cuba and India prove that LC3 reduces CO_2 emissions by 25–35 percent during manufacture compared to conventional cements. This is due to the high substitution of clinker by carbon neutral materials. Preliminary economic studies prove that LC3 cement can be produced at costs 15 percent lower than traditional cements. No other known "green" technology has been able to make such great carbon savings and also be better economically.

We have a material that outperforms current cement with lower emissions and marginal cost savings — great! We still need, however, to address the issues introduced by this new technology. The cement industry worldwide is very conservative; long periods of return on investment, the many substantial risks of problems or failures, and high capital costs slow innovation in (or implementation of) critical new technologies. These barriers to innovation at scale are greatly reduced when the new technology doesn't require substantial infrastructural change. And this may prove to be the real promise of LC3: it already fits the industry as it is.

Kaolinitic soils having relatively low kaolinite content are abundant, but are usually considered *overburden* — the top layer of a quarry that must be removed. The purer layers underneath are richer in kaolinite as used in industry and for the manufacture of pottery. Hundreds of millions of tons of this overburden material, with properties that make it suitable to produce calcined clays, are often stockpiled in abandoned mines and quarries. In just one abandoned quarry in China, there are ten billion tons of overburden material, mine tailings whose composition qualifies it for the production of LC3.

About a third of limestone reserves are not suitable for the production of cement because they have magnesium in their composition (known as *dolomite*), and when fired at high temperatures, potentially hazardous products build up which compromise the concrete's performance and durability. Since LC3 technology does not require firing limestone, these abundant resources can be revalued and used in the production of LC3. Both limestone and kaolinitic clays have a very broad geographic distribution that guarantees that LC3 can be practically produced in any country or region.

The main components of LC3, limestone and clay, are also the main raw materials for the production of conventional Portland cement. Normally a cement plant will have a clay deposit and a limestone quarry at reasonable distance from the plant, so transportation costs and carbon emissions associated with transport will not increase.

The technology for clay calcination is very well-known. There are several kilns in commercial operation producing calcined clay in Brazil and India. Modern technology with higher energy efficiencies such as flash calcination or fluidized bed boilers has emerged, but even in the worst-case, capital expenditures (CAPEX) for new investment is a tenth of that required for clinker production. Further, in almost any cement plant, there is an old, redundant clinker kiln that is left operational to handle contingencies. This kiln can be refurbished for a rather low cost and converted into a clay calciner. Preliminary investment studies confirm that growth of production through the introduction of LC3 could be an attractive alternative in terms of CAPEX and return on investment.

The next barrier would be that cement, as a commodity, should comply with international standards that constrain its composition. The offset of 15 percent of clinker achieved by LC3 in comparison with current pozzolanic cements is not yet accepted in some standards, for instance in Europe, where minimum clinker content is set at 65 percent. Fortunately, the widely accepted ASTM International standard ASTM C595/595M-14 allows a lower minimum clinker content of 40 percent; a ternary blend like LC3 falls within this range. Cuban standardization bodies have taken the lead and have expanded current cement standard to cover LC3 composition, and other standards are expected within the next few years.

The next challenge is the construction industry, the main cement consumer. This industry is very reluctant to adopt "greener" cements with lower clinker content for several reasons. Often the initial strength of concrete is compromised during the first few days due to the slow pace of the pozzolanic reaction. Also, many new cements require more water for mixing, and then require more chemical admixtures to counteract the weakening effect of the water. (Concrete strength varies inversely with water content — drier mixes are stronger mixes.) Further, the reddish color in LC3 due to the presence of high iron clay can be a concern for some clients.

However, real industrial applications of LC3 concrete have demonstrated that it does not require any special care. In less than three days, concrete made with LC3 matches concrete made with pure Portland cement, thus ruling out the issue of early strength. Preliminary durability studies have proven that concrete made with LC3 is three or four times more impermeable than concrete made with pure Portland cement. This should be attractive for construction under severe conditions such as coastal regions with salty air and water, where avoiding corrosion of reinforcing steel is crucial to the service life of the structure.

LC3 is a promising alternative to enable emerging economies such as those in Asia, Africa, and South America — where most of the world's growth in the next 30 years is expected — to choose a green development path and build up their infrastructure with far less cost and carbon emissions. Further, alternatives such as Carbon Capture and Storage, advocated by the Cement Sustainability Initiative (CSI), could be bypassed. Estimates for 2050 show that LC3 technology has the potential of globally reducing up to 400 million tons CO_2 per year, about one percent of global carbon emissions at current rates. LC3 could change the way cement is made and perceived, and help buildings mitigate climate disruption. ■

More Ways To Reinvent Concrete

Many approaches to decarbonizing concrete seek to keep the enormous, complex, and expensive infrastructure already around Portland cement in place, but then offset its carbon emissions. The cement industry has already modernized and made production much more efficient, but there's only so much you can do when your process requires baking huge quantities of rock to very high temperatures. One approach — CarbonCure — is to "supercure" concrete products like blocks and precast panels by exposure to liquid CO_2 captured from cement plant emissions, essentially speeding up and magnifying the natural "soaking up" of atmospheric carbon that all concrete does over time.[3]

Another concrete innovation is borrowed from antiquity: compaction. Part of the secret to the longevity of historic Roman concrete, as well as historic rammed earth structures

Fig. 6.7: *Starting with the same ancient technology — ramming — that gave us the Pantheon and the Great Wall of China (among many others), Watershed Materials makes its blocks with different sources of regional aggregate, then binds them with minimal cement to get strength and durability without the climate penalty.* Credit: © Jacob Snavely

all over the world, is density. Whether you're binding with clay, as in rammed earth, or lime-pozzolan cement, like the Romans, or even modern Portland cement, your concrete strength will increase with density because at the microscopic scale the binder doesn't have to "stretch" so far to hold the aggregate together. To get higher density, you use less water, which also contributes strength, and then mechanically compact the damp mix. That's how the Romans made their famous concrete, and that's how rammed earth is made. The modern concrete industry recognizes this as *roller-compacted concrete*, as is used for massive unreinforced structures like dams, and also *compacted soil-cement*, as is common for road subgrades. Builder David Easton has been building and popularizing rammed earth in California and beyond for decades, adding Portland cement sometimes only to mollify nervous engineers and building officials. The strength of his compacted soil-cement (as it would be categorized in modern parlance) is impressive, for unlike ordinary concretes, it continues gaining strength for years and even decades. The process makes for beautiful walls of artificial sedimentary rock, but is quite labor-intensive and therefore expensive. Seeking to make it affordable, Easton applied the technology to a common product: the lowly but enormously useful concrete block, or cement masonry unit (CMU). His company, Watershed Materials, now makes compacted low-carbon blocks that can be used to replace CMU in many applications with both a better look and a much lower carbon footprint.

Yet another novel innovation, pioneered by companies such as Blue Planet and Calera, is to capture gaseous carbon emissions at cement, power, and metal plants and bind them as artificial limestone aggregate — carbon sequestering sand and gravel. This holds appeal for the carbon storage value, but also for the possibility of relieving pressure on quarries and riverbeds (as noted in earlier mention of "sand mafias") where we currently harvest sand and gravel. It's not a particularly sexy idea, making gravel, but the possibilities are huge: transform widespread industrial carbon emissions into limestone sand and gravel for our concrete, that is needed everywhere — buildings made of sky!

What About Reinforcing — Steel and More

Concrete, whether clay or Portland cement based, is relatively weak in tension. That's why historic structures like the cob apartments of Shibam or the Pantheon relied on massively thick walls at the base tapering to narrow walls at the top; you mimic some version of a pyramid and cross brace or buttress every wall so as to minimize any tensile stresses. Even then, concrete and masonry structures don't last too long in seismically active areas unless they're particularly small and compact or they're reinforced with something strong in tension. That notion dawned on builders at least 1,500 years ago, and early domed structures like Hagia Sophia in modern-day Istanbul used logs embedded in mortaring as tensile reinforcement. Which worked, but not always particularly well.

Enter steel. Basically a modification of iron to get greater strength and, especially, less brittleness, steel appeared in human culture separately in Anatolia, Africa, China, and India as much as 4,000 years ago. Made possible as we got better at harnessing and focusing the energy of fire, wind, and slaves, this early high-carbon material opened up many possibilities (though for many centuries, we mostly used it for swords to kill each other, or hinges and bolts to lock each other out; the cars and refrigerators came much later).

After some fits and starts, reinforced concrete fully arrived a century ago with the first multiple-story concrete buildings in the UK and USA. The art, science, and engineering of reinforced concrete — no simple subject — took off from there and continues to be refined and expanded today. For its nearly limitless fluidity and malleability, it is catnip to architects, but is still rooted in the two most carbon-intense materials we work with: steel and Portland cement. Aside from the fact that steel reinforcement often ends up as the downfall of a concrete structure, usually by rusting and spalling, we have to ask in this book's context if we have lower-carbon reinforcing alternatives.

Yes, and no. First, distinguish *local* from *global* reinforcing. Local reinforcing strengthens just a part of the building (straw fibers in a clay plaster to reduce cracking and erosion, or steel fibers in the "knuckle" of a concrete structure that add toughness), while global reinforcing resists bending and tension in key members like beams and columns, and gives continuity to the entire structure as in foundations or bond beams. Lots of things can be used for local reinforcing, not so many work globally.

The current stable of alternatives to steel as global reinforcing are few. Bamboo has been tried and studied extensively, as it is very low carbon, grows quickly, and has remarkable tensile strength (around 15,000 pounds per square inch). Its limitations as concrete reinforcing, however, constrain it to only the simplest uses; it won't bend around a corner, doesn't bond well to the surrounding concrete, has a relatively low elastic modulus, and can decay. Fiberglass, carbon fiber, ceramics, and basalt (melted igneous rock) rebar are around, but each is either higher in embodied carbon and/or is much harder to work with; none show promise as low-carbon replacements for steel.

The steel industry writ large — rebar, wide flange shapes, cars and trucks, appliances, etc. — already recycles at a very high rate. This is not born out of noble efforts to preserve a stable climate, but simple economics: it's increasingly cheaper and easier to harvest used steel from human industry than from ore in the ground (see Chapter Two for a discussion of the embodied energy/carbon of steel). We have as a species already pretty much found and extracted all the high assay reserves of most metals in the world. Copper, for example, was discovered thousands of years ago on the island Cyprus that shares its name, because it occurred there in nearly pure metallic form and was easy to study and harvest. Now, global industries will tear up mountainsides to harvest copper that is but a fraction of a percent of the mined earth, usually to the detriment of local ecosystems and human culture.

So, to a very large extent, and for now, steel is it. From a climate perspective, it's good that steel is so highly recycled and recyclable, but maybe not so good that it still requires intense energy input to collect, sort, melt, and reuse it; we're not seeing any solar steel mills coming along anytime soon. As a practicing structural engineer, I can say that, through 40 years of practice, I have consistently seen concrete made with too much cement and too much steel — more by far than really needed. Decarbonizing reinforced concrete will surely involve some reinvention of cements and reinforcing, but also training and incentivizing engineers, building regulators, and builders to be more careful in their use of high-carbon materials.

Two hundred years ago, when Portland cement was invented, sand, gravel, and fossil fuels were cheap and abundant, and there was no problem with carbon emissions. Today those conditions no longer hold, yet we continue to make concrete — by far the most common building material in the world — in pretty much the same way. We also continue to be fruitful and multiply, so it's time for change. Change doesn't come quickly in this industry, and for good reason, but as each day passes, the barriers to change are not so much technical as cultural: Portland cement and concrete are deeply wedded in people's minds, but now need a divorce, or at least a bit of separation. Like the buildings and cities that use it, concrete can and must switch from climate villain to climate champion.

Notes

1. www.slideshare.net/smeseguer/the-origin-of-clay-minerals-1-1
2. "Calcining," or *calcination*, is a fancy way of saying *heated*. Calcined clay or rock has been baked, to one degree or another, so as to change its chemical or mechanical properties.
3. When exposed to air, the outer inch or so of hardened concrete very slowly absorbs CO_2 in a process called *carbonation*. As it occurs naturally, carbonation is not thought to be an important factor in concrete's carbon footprint.

Chapter Seven

Plastic: So Great, So Awful — Some New Directions

Mikhail Davis, Wes Sullens, and Wil Srubar

I'm stubborn as those garbage bags that time cannot decay.

— Leonard Cohen

Introduction

P LASTIC IS FANTASTIC. You can use it to hold liquids, shed rain, insulate wires and roofs, seal leaks, store food — the list goes on and on. In fact, you've probably got one or more forms of plastic on your body; it is what we wear. You probably ate breakfast today that, at least partly, was grown, fertilized, harvested, processed, transported, refrigerated, delivered, and prepared with petrochemicals and plastics; it is what we eat. And there are most certainly dozens or hundreds of plastics in the building around you; it is where we live. We use it for all these

Editor's Introduction

Plastics hold an odd place in our discussion. They are everywhere in our material economy and so need some attention here, but, though comprised of carbon, they don't show any clear promise as a carbon-sequestering strategy within the context of this book. They do, however, show a clear and enormous threat to the health of ourselves, our descendants, and the systems of life all around us; we are only now discerning that almost every bit of plastic we ever made is still here, somewhere, as Nature has evolved no means of eating or recycling it. It lives on as heart-rending chokers around marine mammals, as lethal gut stuffing in birds and plankton, as trash in the landscape, and as toxic, smoky fuel for the cook fires of millions of urban poor. The purpose of this book is to promote pulling carbon from the air and into buildings so as to have a healthier climate; including a look at plastics just makes sense in seeking ways to build that don't cause so much harm.

things because it does them all so much better than anything we ever had before.

And that's the trouble: we depend on plastic every day in hundreds of ways. As terrific as plastic is, it can be equally terrible. Non-biodegradable plastics choke ocean life of every size, from plankton to birds to whales. Its disposal is a worldwide quandary, for none of nature's living systems — evolved as organic life over billions of years — know how to deal with it, since it is a synthetic (human-made) material. Its production releases a nightmarish panoply of chemicals into air, soil, and water, while the science to understand those chemicals and the rules to control them lag far, far behind our frantic race to use ever more in ever more forms. Just as the use of fossil fuels has been an irresistible source of cheap, intense energy, the use of petrochemical plastic compounds has proven to be a fabulous, bountiful gift whose hidden costs have only recently come into view.

So what's to be done with this too-useful, fossil-fuel-derived material that degrades only after decades or even centuries in the environment? What role might plastics play in a beyond-zero future where buildings become part of the solution to climate change and other pressing global problems? What technological solutions are being developed to help make a brighter future for plastics and the environment?

In this chapter we explore some of these questions and explain and evaluate the potential paths and barriers to how plastic might actually redeem itself and prove that it has a meaningful role in the regenerative future of our built environment.

Biopolymers and Bioplastics

Polymers and plastics can be chemically described, simply and endearingly, as HOOCH-y COOCH-y materials — they are made from hydrogen (H), oxygen (O), and (you guessed it) carbon (C). Environmental issues around modern plastics stem from the source of the carbon: petroleum — a cheap and historically plentiful source of the hydrocarbons that are the building blocks of many industrial plastics like polypropylene and polyethylene. This carbon is termed *abiotic* or *non-biogenic* carbon — carbon from non-biological sources. (Petroleum comes from the fossils of plants, tiny marine organisms, and larger animals; thus the term *fossil fuels*. But abiotic carbon is, in effect, carbon that is nonrenewable on the human time scale.) Contrastingly, there exist abundant sources of *biotic* or *biogenic* carbon that can be used to make biopolymers and bioplastics, such as plants, animals, and even bacteria.

Plant Biopolymers

We have been building for millennia with a bio-based polymer called wood, which is comprised of three main components: cellulose, hemicellulose, and lignin. Cellulose, the most abundant carbon compound on the planet, is highly crystalline and gives structure to wood and other natural fibers, while hemicellulose is a less ordered, less strong form of cellulose, and lignin is the glue that holds it all together.

Today, we are building ever taller and larger structures of wood that rival the heights of steel and concrete structures (see Chapter Four: Wood). Wood is also biodegradable and requires maintenance, but with care will last far beyond the conventional 60-year design service lives of modern buildings. Some notable examples include Japanese pagodas built in the 7th and 8th centuries, which have withstood countless earthquakes, and timber-framed structures all over Europe, which have endured freezing and wet weather for centuries.

Chemically, cellulose and the starches also found in plants are *polysaccharides* — sugar. Their different arrangements are what give plants strength (cellulose) and energy (starch) — and also why we as humans can digest one and not the other. The different chemical arrangement is also why manufacturers find it easier to convert starch to usable biopolymers and bioplastics rather than cellulose.

Starches provide a rich source of C, H, and O that can be used as a building block for biopolymers that resemble conventional plastics. One of the most well-known global producers of plant-based bioplastics is NatureWorks, which produces a plastic called polylactic acid (PLA) from sugars derived from corn starch and sugarcane. Products made from PLA are now in the market, the most common being utensils, biodegradable packaging materials, and compostable plastic cups. However, while PLA is technically compostable, there is no official recycling stream that can handle PLA plastic; it requires municipal-scale compost equipment.

Aside from sugars and starches, other biopolymers exist in all cellular life — arguably the most famous is deoxyribonucleic acid (DNA) — as well as other natural proteins.

Plant protein is one of the major sources of biopolymers on the planet, and many plastics have been made from it, the most economically competitive being soy. Soy plastics were heavily researched in the 1920–1940s and were used to decrease the cost of petroleum-based plastics and enhance their biodegradability. Henry Ford researched soy in the 1930s and patented a myriad of soy-based technologies for his Model A cars and Fordson tractors.

Soy-based plastics are made from soybean oil and are used in the manufacture of sealants, adhesives, coatings, foams, and insulation. If combined with the right chemicals (like isocyanates), soy-based products can rival petroleum-based products in strength, durability — and non-biodegradability. This tradeoff demonstrates that, while some polymers and plastics can be derived from biorenewable resources, the actual chemical processing into a usable plastic may negate end-of-life environmental benefits.

Animal Biopolymers

Animals provide us with many nonfood products such as wool, silk, leather, and feathers. But they are also an abundant source of proteins, like casein from milk and gelatin from skin and bone, both of which have remarkable adhesive properties.

Historically, adhesives derived from animal proteins were used in wood composites until their replacement by synthetic, low-cost, higher-performing ones. In addition to advantages such as biorenewability, biodegradability, global availability, and non-toxicity, gelatin films exhibit excellent adhesive properties compared to other biopolymers and are now in use as a bioadhesive for engineered wood products.

About 60 percent of all adhesives manufactured worldwide are used in the fabrication of engineered wood products. Conventional adhesives are made using phenol formaldehyde, urea formaldehyde, or polymer isocyanates — materials identified as either allergens or carcinogens. In addition to health concerns, the use of adhesives is problematic because it impedes future reuse and deconstruction, as when floor sheathing glued to joists makes the assembly quieter but difficult to take apart or change; design for adaptability is a fundamental part of climate-friendly architecture.

Bacterial Biopolymers

Microbes (microorganisms, such as bacteria) are fascinating — and everywhere. Some firms have begun to produce bioplastic using bacteria combined with waste methane gas (a dense carbon source) from a wastewater treatment plant or a landfill. The result is a biodegradable polymer, namely polyhydroxyalkanoates (PHAs). Once that product is no longer being used, it can be sent to a wastewater treatment plant or landfill to be degraded and turned back into methane. One drawback of PHAs is that they are very highly crystalline and brittle, but those properties can be tailored to make blends with desirable characteristics.

The Bioplastics Dilemma

The thermal properties of most biopolymers are not up to snuff with petrochemical polymers, not only because the plastic is a poor insulator, but also because it has a low *glass transition temperature* — it will flow and permanently distort.

Another issue arises with gelatin-based adhesives: gelatin is hydrophilic, meaning that it loves water. So while it is dry, it is an excellent adhesive, but not so when wet. This bio-based/durability conflict is also the reason why soy-based plastics are mixed with other chemicals like isocyanate to make them more resilient — and the reason that we must paint or varnish wood to keep it from fading and degrading in rain and sun.

This illustrates a simple and important tradeoff with most bio-based building materials. They are certainly biorenewable, bio degradable, and perhaps even compostable — all excellent qualities for the sustainable- and carbon-minded architect or engineer. However, we want our materials to be environmentally benign — and also last a really long time.

How do we strike such a balance? One solution is cultural: a pragmatic shift in our relationship with materiality. Should we abandon the desire for long-term durability and instead care for our materials as we do a living thing? Should our ancestors' rituals of maintenance and constant refurbishment re-enter our regular routine? Or should we continue to engineer our biomaterials to be as hyperdurable as their synthetic counterparts? People don't tend to change their ways unless offered something cheaper, easier, or more fun, and there is no simple answer here. Still, these are pertinent questions to ask about *what* we use, *how* it is made, *where* we use it, and *how* long we intend it to last. After all, there's an awful lot of plastic already in existence. Can we find a way to reuse our world's existing plastic over and over while we innovate and transition to bio-based solutions?

Existing Plastics in the World

Petrochemical plastics have been a disruptive technology since they became increasingly mainstream following World War II. They started out replacing products made of metal, wood, or fabric because they were lighter, more flexible, stronger, and cheaper than the natural materials they replaced. Today, plastics have become so advanced that they can replace or be combined with nearly every natural material to improve its characteristics, and also occupy an entire new class of man-made products that were inconceivable before the plastics revolution. This success has led to a mind-boggling array of complicated monomers, polymers, and heterogeneous materials as can now be found everywhere in the built environment. These advances are not without problems, especially when those fabulous plastic products reach the end of their use and are discarded as waste.

The Scale of the Plastics Problem: How Much Is Already Out There?

Since 1964, global plastics production has increased 20-fold, reaching a staggering total of 340 million US tons of plastics in existence in 2015. That amount of plastic weighs as much as a trail of elephants walking in a line, trunk to tail, that would circle the Earth 19 times.

This amazing growth of plastics production is expected to accelerate; some experts estimate plastics production will double again

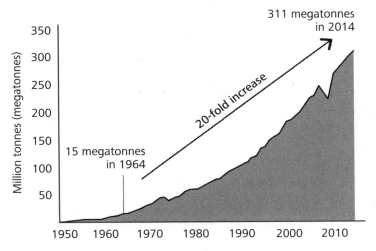

Growth in global plastics production, 1950-2014

Fig. **7.1**: *Based on PlasticsEurope: Plastics — the Facts 2013 and 2015. Note: production from virgin fossil-based feedstock only. (Does not include bio-based, greenhouse gas-based, or recycled feedstock.)*

in the next 20 years and almost quadruple by 2050. At that rate, the *annual* production of plastics post-2050 would equal the total amount of all plastic that existed in 2015 (put another way, that's a trail of elephants circling the Earth 19-wide put into the biosphere *every year*). Currently, plastics already end up as pollution in our environment in staggering amounts.

Analysis done for the New Plastics Economy initiative estimates that by 2050 the amount of plastics in the world's oceans from dumping and runoff could exceed the weight of all fish in the seas — which is not surprising, considering all the single-use plastics that are made every day. For example, packaging materials (candy bar wrappers, pop bottles, take-out containers, etc.) have a pretty dismal fate globally: 40 percent of all plastics packaging are landfilled, 32 percent are leaked to the environment as pollution, and 14 percent are burned for volume reduction or energy generation. Only 14 percent of plastics packaging are collected for recycling currently.[1] While packaging is the number one source of plastic use globally, the second largest is the building sector. In Europe, the building and construction sector are responsible for nearly 20% of annual plastic use.

What To Do With All That Existing Plastic?

Humans do a lousy job of managing plastic; in the US, less than 10 percent of plastic in the waste stream gets recycled. While some plastics like PET bottles have higher recycling rates of 30+ percent, the majority are either landfilled or incinerated rather than turned back into useful products through recycling. Most of the focus on plastics recycling is centered on packaging materials because they are high volume and easily collected curbside. Building materials — with their complex formulas, long life spans, and lack of collection infrastructure — have much lower recycling rates.

Barriers to Plastics Recovery and Recycling

1. Reusing plastic is not easy.

We lack collection systems, places that take back or repair plastic parts, and furthermore plastic degrades over time; plastic doesn't remain plastic. Since the beginning of the plastics heyday, plastic products have been seen as economical alternatives to other costly materials, and thus are typically discarded after one use or installation. Plastic building materials are not made to be taken apart or reused; your plastic decking or laminate flooring, for example, can at best be ground up into particles for use as a low-grade filler.

2. Recycling plastics is not easy.

A logo with chasing arrows may indicate the potential for recycling, but that's a far cry from a plastic actually being recycled. All sorts of problems arise. Is the plastic resin homogenous? Is it glued or attached to something else? Are there labels on the package or additives that need to be removed so that it does not contaminate the next use of that plastic? Unfortunately, recycled plastics as a

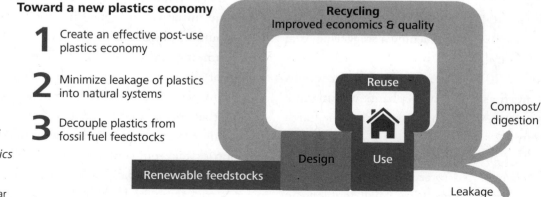

Fig. 7.2:
Aspirations of the New Plastics Economy.

Credit: Wil Srubar

Type	Recycling code	Common names	Common building products	Recycled content?
Polyethylene (PET)	♻ 1 PET	PE, LDPE, LLDPE, HDPE, polyester	Pipes, decking, lumber, partitions, substrates, rigid board, wall and ceiling panels, panels, tabletops, carpets, fabrics	Decking, lumber partitions, panels, carpets
Polyvinyl chloride (PVC)	♻ 3 PVC	Vinyl	Window frames, pipes, flooring, roofing membranes, wires and cables	Flooring
Polypropylene (PP)	♻ 5 PP		Carpeting, thermoplastic fiber reinforced composites, fabric, ropes, packaging	
Polystyrene (PS)	♻ 6 PS	EPS, XPS	Appliances, electronics, insulation, packaging, roadbank stabilization systems	
Polyurethane (PU)	♻ 7 Other	FPF	Paints, varnishes, adhesives foams (furniture, carpet pads, mattresses), insulation, composite wood panels	Carpet pad
Polyamides	♻ 7 Other	Nylon	Carpets, drapes, bedding, outdoor fabrics, flooring, molded products	Carpet, flooring

Fig. 7.3: *While convenient, the "chasing arrows" don't really tell you about the additives or other things in a plastic, and don't necessarily indicate recyclability.* Credit: Wil Srubar

product are not as clean, neat, or uniform as the virgin plastic products they are trying to replace, and that means using recycled products takes more effort than making things new. The recycling industry has gotten good at recycling some plastics with fancy conveyor belt sorting systems that have air jets and optical resin sorting lasers. Yet those hi-tech sorting lines require significant investment and a constant supply. For the most part, plastics from the building sector are not sorted in such ways.

3. What's in all that plastic?

Another complicating factor for plastics recycling is the myriad additives that find their way into different products. These additives give the plastic qualities we want like color, opaqueness, flexibility, strength, and more, but also cause problems. The process of manufacturing plastics introduces numerous chemicals, some of which are hazardous or toxic during the manufacturing and/or use phase. Furthermore, these plastics can make their way into the recycling streams with sometimes decades of delay, which can lead to products that are now banned being reintroduced into new products. An example is PVC flooring: a consumer watchdog group tested samples of resilient PVC floor-ing with recycled content in 2015 and found high levels of lead and phthalates in interi-or layers, most likely present because of the recycled cable scrap wire that was ground up and used in the flooring. The cable scrap came from recycled wire like wiring and elec-trical cabling that were up to 40 years old.

The non-profit BizNGO published a Plastics Scorecard that ranks the relative sustainability and inherently hazardous feedstocks of various plastics. The Scorecard plastic polymers today, and also provides

recommendations for reducing the chemical "footprint" of plastics. The graphic below shows an overview of findings, and begs the question of readers: How do we move these plastics upward and toward the right?

4. Economics of plastics recycling

The economics of the recycled plastics com-modity markets are fickle, at best. The building blocks of most plastics are fossil fuels: oil and natural gas. In recent years, a glut of domestic natural gas production combined with new oil reserves and more economical extraction

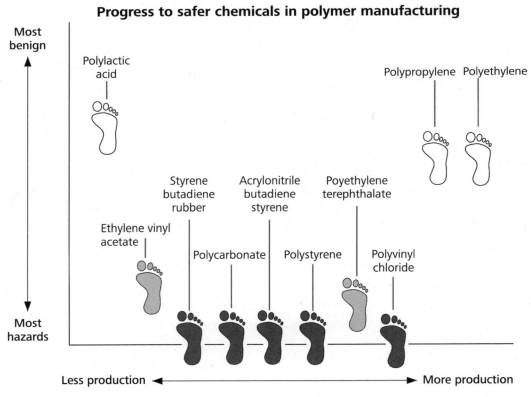

Progress to safer chemicals in polymer manufacturing

Fig. 7.4:
*Hazard
"footprints"
of common
plastics: based
on the Plastics
Scorecard.*

methods like fracking has driven down the cost of raw materials. The plastics industry has benefited greatly from this outcome, while the recycling markets have suffered; recycling many forms of plastic costs more than buying virgin materials. Recycling involves collecting scrap materials, sorting them out, preprocessing so they can be used again, and then manufacturing them into a new product; when oil is cheap, recycling rates plummet. Unless plastics recycling is subsidized, and the price for virgin plastic commodities reflects the full life cycle costs of plastics production, pollution, and end-of-life handling[2], recycled plastics can't compete with clean, new molecules created from cheap oil and gas. (See also Chapter Eleven on carbon pricing.)

Bright Spots for Plastics

Plastics are a miracle technology, and we will continue to want and use them for a long time, with all their many warts and smells. But all is not doom and gloom.

Global initiatives like the New Plastics Economy (www.newplasticseconomy.org) are setting ambitious private-public partnerships between government and industry to help alleviate pollution and create closed-loop systems where recycled plastics become feedstock. In the green building space, whole building rating systems like LEED Version 4 by the US Green Building Council and the Living Building Challenge by the International Living Futures Institute reward project designers that select products with lower intrinsic hazards because they have screened out the worst chemical offenders. Product certification schemes like Cradle to Cradle and UL Environment's Product Lens are used by progressive manufacturers that have studied the ingredients in their products and analyzed potential exposure pathways for humans and the environment. As knowledge grows from these and other efforts of transparency, the industry will gain a greater understanding of what is in products, and as these are disclosed, consumers and designers can make better choices today that will affect the ability to cycle those products in the future.

In the recycled plastics space, StopWaste of Alameda County, California, partnered with Healthy Building Network in 2015 to develop a series of studies seeking to optimize recycled content in building materials including PVC, polyethylene, polyurethane, and nylon plastics (see www.healthybuilding. net/content/optimize-recycling). This effort analyzes recycled materials in four categories: environmental impact, supply chain health and transparency, potential for green jobs creation, and scalability ("room to grow").

What You Can Do: The Low-carbon Plastics Hierarchy

Solving the problems of plastics will take a multifaceted effort that goes beyond just buying recycled or selecting a bio-based plastics. What's needed are collaborative agreements and commitments by building owners, governments, manufacturers, architects, and contractors to help close the loop on plastics in the built environment. There are efforts underway to achieve such outcomes, but they need your help. While the actions of one individual project may not

Post-consumer recycled content feedstock evaluation summary

Post-consumer recycled content feedstock	Environmental & health impacts	Supply chain	Green jobs	Room to grow
Nylon 6 scrap	●	●	●	○
Glass cullet	●	○	●	●
Polyethylene scrap	●	○	○	●
Recycled asphalt shingles	○	○	●	●
Nylon 6, 6 scrap	○	○	●	○
Reclaimed asphalt pavement	○	◉	●	○
Ground rubber (from tire scrap)	◉	○	○	○
Recycled wood fiber	◉	◉	○	●
Steel scrap	◉	◉	○	○
Flexible polyurethane foam scrap	◉	◉	○	○
Polyvinyl chloride scrap	◉	◉	◉	○

● Very good   ○ Room for improvement   ◉ Significant concerns

Some of these feedstocks can be obtained from clean sources and will result in a very good (or " ● ") score.

Some of these feedstocks can be tested, processed, or screened via identified best practices that will result in a very good (or " ● ") score.

Fig. 7.5: *Healthy Building Network and StopWaste produced the Optimizing Recycling Report Series which investigates common recycled content materials found in building materials, and evaluates them on four environmental criteria.*

seem like they can make a difference, they do. For those who seek to follow best practices, we recommend the following guidelines so as to have the greatest impact possible.

Guidelines: The Low-carbon Plastics Hierarchy

1. **Use things that already exist.** Rather than create new molecules — no matter how benign those new monomers and polymers may be — let's make use of the stuff we've already created. These should be a resource, not a legacy. Take them out of the oceans, out of the trash bin, or re-purpose them. Get creative.

2. **If already-made plastics are toxic, seek solutions that don't expose humans or the environment.** Tainted plastics may have some use in downcycled, encapsulated, or other products and applications where they are protected from exposure to humans or releases to the environment.

When that's not possible, choose the lesser of evils and prefer disposal options that protect the environment from releasing those contaminants.

3. **Use plants, not dead dinosaurs.** If new plastics are unavoidable, choose products that come from bio-based sources instead of fossil-fuel-based raw materials. Prefer manufacturers that source their supply from sustainably harvested sources, such as the Sustainable Agricultural Network or USDA organic.

4. **If you have to buy new plastic, choose clean.** When it absolutely has to be new, prefer products that are free from hazardous chemical additives. Ask manufacturers what's in them and demand transparency reports that are independently verified. Clean plastics have a much greater chance of being perpetually recycled safely, and that's what we should be buying from today forward.

5. **Close the loop.** Buy from manufacturers that take back their scraps or used products and reuse or recycle them.

6. **Know your plastic's footprint.** With Environmental Product Declarations (EPDs) common in the building industry today, there's no reason you shouldn't be able to look up the Global Warming Potential of the plastic product you're buying.

7. **Celebrate and share.** Any success, however incremental, fosters global environmental improvement. We encourage those who make strides toward a low-carbon plastics economy to celebrate their results and share them widely.

From Obstacles to Opportunities to Solutions: Can We Redeem Plastic?

We have looked at opportunities and obstacles that plastics must overcome to become part of a regenerative future for architecture. Here's what we need.

Performance: Make recycled or bio-based plastics just as useful as the petroplastics we have come to rely on. Make them sufficiently durable, heat-resistant, flexible, etc. or offer performance trade-offs that most users will find acceptable.

Economics: Make recycled or bio-based plastics competitive on price with petroplastics regardless of the challenges of efficient recycling and/or bio-based production and the current subsidies granted to the petrochemical industry. (Or just change the subsidies; see Chapter Eleven.)

Scalability: Move production of better plastics from small scale to large scale without causing new problems, and spur research and investment in new materials.

Circularity: Many plastics are theoretically "recyclable," but is it really possible to create systems that can keep plastics cycling in perpetuity in the economy? We need to find a way.

On the flip side, we see many hopeful potential opportunities:

1. Trash to Treasure: Can we harvest the existing plastic pollution from the environment to make new products?
2. Carbon-capturing Materials: Can we produce plastics that capture and store carbon?

3. Closing the Loop: Can we truly manage plastics in a circular system?

Keeping in mind the sage words of David Brower, "It's far too late and things are far too bad for us to indulge in pessimism," we will use a series of micro-case studies to explore the current progress in turning each of the opportunities into a reality.

Trash to Treasure: Can We Harvest the Existing Plastic Pollution from the Environment to Make New Products?

Yes, but there is a steep learning curve for companies looking to incorporate these cast-off materials into their supply chains, materials, or packaging. Method Home, San Francisco-based maker of cleaning and personal care products, put years of effort into trying to make a soap bottle from ocean plastics, finally concluding that the plastic floating in the ocean was too mixed, contaminated, and degraded by sunlight to ever make a structurally sound bottle. They have switched their efforts to collecting plastics from beach cleanups, capturing bottles and other plastic litter just before they permanently become ocean plastics. There have been much publicized technologies proposed to capture plastics floating in the open ocean, but they don't offer much in the way of reuse options. The most effective efforts to date have been focused on prevention and nearshore capture (beaches, harbors, bays, etc.) of plastic pollution before it reaches the open ocean.

For example, global carpet tile manufacturer Interface partnered with Aquafil, a nylon yarn manufacturer, and the Zoological Society of London (ZSL), a global conservation organization, to prove out a model for successfully harnessing ocean plastics in buildings in a way that provides environmental, social, and economics benefits. This initiative is called Net-Works.

After designing the pilot program with Interface and Aquafil, ZSL and local NGO partners began by setting up local community-run savings banks in fishing villages in the central Philippines and offering a per-kilo price for old nylon 6 fishing nets turned in to the bank by local residents. This immediately led to widespread efforts to harvest and clean the nets befouling the local shoreline, docks, and reefs. It also created an economic incentive for local fishers (most of whom live well below the poverty line) to turn in their damaged nets, rather than dumping them on the beach or in the water. Communities that have the Net-Works program stand out in a region where most other beaches are littered with nets.

Local ZSL representatives ensure that nets are packed for international shipping to Aquafil, whose EcoNyl Regeneration technology produces a 100 percent recycled nylon that has no residual additives and performs like virgin nylon. This recycled nylon from multiple post-consumer and pre-consumer sources has a carbon footprint at least 50 percent lower than virgin nylon 6 made from petroleum.

Today, Net-Works operates in over 30 communities in the Philippines and has set up banking and net collection in several fishing-dependent communities in West Africa.

Over 150 metric tons of nets have been collected for Aquafil to turn into carpet yarn for Interface carpet tile products.

But what about scalability? Net-Works represents a small fraction of the fishing nets and other nylon waste that Aquafil is converting into EcoNyl yarn. Though some more heroic efforts by the Healthy Seas Initiative have sent divers down to harvest "ghost nets" in the open ocean (51 metric tons and counting), the vast majority of the post-consumer materials in EcoNyl come from the collection of old industrial fishing nets in major ports or of nylon sheared from old carpet products. While EcoNyl yarn remains standard in the majority of Interface products globally, it is also being used now by many other carpet and apparel manufacturers.

Note that nylon, because it is an engineering plastic suitable for replacing metal in its strength and durability, is more expensive than many common plastics like polyester or polyethylene. The fact that nylon fishing nets are one of the most expensive forms of ocean plastic is the key to making Net-Works and other nylon recycling possible. The fact that nylon nets are uniquely dangerous in their ability to entrap wildlife and that virgin nylon has one of the highest carbon footprints among commonly used plastics has not yet been a factor.

(Add to our wish list : "Plastic magnets" that every transoceanic ship is required to drag behind, converting their plastic harvest into clean synthetic fuel to power their voyage. Get your engineer friends working on it!)

Carbon-loving Plastics: Can We Produce Plastics that Capture or Store Carbon?

Interface announced in June 2016 that they were raising the stakes on their more than 20-year-old mission to eliminate any negative impact the company has on the environment by 2020. Their new Climate Take Back mission is focused on reversing global warming by operating the company in a way that removes carbon from the atmosphere. This includes a new mandate to "Love Carbon." What could it mean for a carpet tile or any other building product to love carbon? Loving carbon means embracing it as a solution, not a problem, and as a building block for raw materials and products. It means finding materials that capture or store carbon and using them to make products.

One of the best ways to reduce the carbon impact of a plastic is to make it from petrochemical wastes we've already created, like fishing nets or plastic bottles, rather than from new fossil carbon pulled up from under the ground. But this "carbon reduction" approach to building materials would be unfamiliar to a tree or a coral reef. Wood or even seashells are not "low-carbon" materials; they are *made* of carbon, directly harvested from air or water. If plastics could learn to love carbon, maybe we could all learn to love plastic again.

Paths to Bio-based Plastics

But don't all bio-based materials "love carbon"? Not necessarily, but at least three promising pathways exist to move beyond mere bio-based to true carbon capturing and storing plastics.

Regenerative Agriculture

Conventional industrial agriculture produces bio-based products, but does so in a way that contributes to global climate change rather than capturing and storing carbon. Most of the feedstocks for bio-plastics today increase the use of energy-intensive fertilizers, runoff of fertilizer to rivers and oceans, and the use of fossil fuels in farm equipment, at least partially negating their carbon benefit. When you factor in the limited bank of agricultural land, none of the crops that compete with food crops for land are truly scalable as plastic feedstocks, even if the agricultural practices can be improved.

The true hope for bio-plastics like polylactic acid (PLA) or bio-PET (Coca-Cola's Plant Bottle initiative), made today from corn or other commodity crops, involves a complete overhaul of agriculture practices, often called *regenerative agriculture*. This takes many forms but always involves building and preserving soil rich in organic matter, which sequesters carbon. Without this change, even a next generation of "cellulosic" (non-food) feedstocks would do little to help the climate.

GHG to Plastic

Another way to for plastic to love carbon would be to skip using the bodies of plants or algae for feedstock and build our own polymers directly from greenhouse gases, mimicking the way a tree would do it.

This is not science fiction: GHG plastics are here but just not at scale yet. Novomer and Covestro have made polyurethanes at least partially from CO_2. Mango Materials makes PHA plastic from waste methane (22 times more powerful than CO_2 as a greenhouse gas) from wastewater treatment plants using microbes, while Newlight Technologies uses enzymes to make a similar PHA ("AirCarbon") that is now used in several lines of KI brand office chairs.

At least a dozen other companies globally have technology to turn CO_2 into various types of polymers, and The Carbon XPrize has put up $20 million for the best new use of all the excess CO_2 in our atmosphere.

Next come the prototype products. Power company NRG partnered to create a "Shoe Without a Footprint" with a plastic sole made from their power plant emissions, and Interface has prototyped flooring products that capture carbon and processes that sequester carbon.

Carbon-plastic Composites: Can We Put New Carbon into Old Plastic?

Going a little deeper, there are now composites in which the plastic portion may not be capturing carbon, but the mineral components are.

- "Carbon-fiber" like we increasingly see in car and airplane bodies, made of plastics mixed with more rigid fibers, can indeed be made with atmospheric carbon, at least in a laboratory.
- Research at the University of Colorado has made wood fiber composites transparent enough that we could see wooden windows someday.
- A huge range of currently petroleum-based plastic additives now have bio-based competitors, including bio-based softeners and

lubricants that could be added to recycled plastics to increase their carbon content.

+ An even more exciting class of plastics additives are fillers, where cheap limestone powder (calcium *carbonate*) is commonly added to a huge range of plastics, up to 80 percent of the weight of some flooring types. Blue Planet Ltd. is commercializing technology that can make artificial limestone from power plant emissions cheaper than mined limestone (see Chapter Six: Concrete), which could transform the carbon footprint of filled plastics from positive to negative with no reformulation required.

Closing the Loop: Can We Truly Manage Plastics in a Circular System?

The question of closing the loop on plastics remains open. We can bemoan unfair economic advantages of virgin plastics and the imperfection of the additives we find in waste plastics, but unless we get serious about closing the loop, plastics cannot hope for redemption. We do not have time to wait until all the plastics are bio-based, CO_2-based, or even 100 percent free of all potentially hazardous residuals to *then* start recycling.

We should look to pioneers like Dr. Michael Biddle, founder of MBA Polymers, for the answers to the problem of creating treasure from actual trash. The company's technology can sift through waste as mixed as "shredder residue" (what's left over after

you take the metal out of a car) and shredded e-waste or appliances, and put out five distinct streams of virgin-quality post-consumer plastics.

And what about multiple life cycles for recycled plastic? Most plastic recycling occurs in a remelt system, when plastic must be heated into a liquid to become a new product. This degrades the plastic a bit each time. Some plastics survive multiple "heat lives" better than others. This is the technical side of plastic recycling, once we get past the collection logistics, the economics, and the additives screening. We still need to pick plastics and recycling technologies that allow plastics to be recycled more than once, or we are only delaying their inevitable return to those places we don't want them, like landfills, incinerators, and the ocean.

What does all of this mean for industries reliant on plastics? If plastics are to be redeemed and play a role in the regenerative future of buildings, many things will have to change, from the expectations we have for building maintenance, to the sources of the carbon in new plastics, to our ability to harvest, sort, and process plastic wastes effectively. Ultimately, the future of plastics will depend on our ability to combine these strategies effectively, to ensure that whatever the source of the plastics we continue to use, they are keeping carbon out of the atmosphere and trash out of the ocean.

Notes

1. *New Plastics Economy: Rethinking the Future of Plastics,* 7. www3.weforum.org/docs/WEF_The_New_Plastics_Economy.pdf

2. *Project Mainstream* estimates that 32 percent of plastic is leaked into the environment as pollution, clogging up storm drains in cities and polluting marine and other natural environments. The report *Valuing Plastic* conservatively estimates the costs of the negative externalities of plastics in the oceans to be at least USD $13 billion. (*New Plastics Economy,* 76.)

Chapter Eight

To Your Health: The Health Benefits and Impacts of Natural Building Materials

by Pete Walker, Andrew Thomson, and Daniel Maskell

PEOPLE IN MORE ECONOMICALLY DEVELOPED countries spend up to 90 percent of their lives indoors, and the quality of the indoor environment has a significant influence on their well-being, productivity, and long-term health. Indoor air can be up to 10 times worse than outdoor air,[1] resulting in up to 1,000 times more contaminants inhaled.[2] Indoor air quality is dependent on humidity and the level of pollutants such as volatile organic compounds (VOCs), carbon dioxide, radon, and particulates. Indoor environmental comfort also includes thermal and acoustic performance. As we will

Editor's Note

In this chapter the authors use but don't define the term "natural materials." I have worked with and in parallel to Professor Walker for many years, researching clay and straw bale construction, and other staples of the natural building movement, but don't think that anyone has ever tried to define the term very precisely. For our purposes, we should try, and here goes. Natural materials are materials harvested directly from the landscape with minimal or zero processing. They often (but not always) derive from historic and indigenous building systems, and are materials your great-great-grandmother would recognize. In economic terms, they are materials with an extremely short and simple supply chain, and therefore a very small carbon footprint; they might (less poetically) also be called very low-carbon materials. Natural materials, for example, include wood, clay, stone, straw, wool, and bamboo. Less obviously, and arguably, there is also a category of "urban natural" that includes material discarded by an industrial society such as tires, newspaper, cotton clothing, bottles, wood palettes, and shipping containers, but these are not part of the discussion that follows.

see below, problems with indoor pollutants are closely linked to thermal comfort levels. Health problems associated with poor air quality include asthma, respiratory infections, and some allergies.

Recent research has suggested that the problem may be getting worse as a result of changes in building design.[3] Drivers for improved energy efficiency have seen levels of airtightness improve so as to minimize uncontrolled heat losses and gains. However, the unintended consequence has been a degradation of indoor air quality with an increase in levels of VOCs and extremes in relative humidity. Modern energy-efficient buildings rely on sophisticated heating, cooling, and ventilation systems that require careful operation and maintenance, but building designs that minimize air leakage have led to problems with indoor environment quality.

The health, productivity, and well-being benefits of using natural building materials have long been championed with much anecdotal evidence.[4] Current research confirms that natural materials contribute to the improvement of the indoor environment.[5] The most widely studied subject is the passive regulation of indoor relative humidity levels through moisture buffering and breathing fabric performance. Relative humidity (RH) is a ratio of the amount of water vapor in a given volume of air — the ratio of the partial pressure of water vapor to the saturated water pressure at a given temperature. RH varies with temperature and pressure; warmer air can hold more water vapor. The health and well-being benefits of natural building materials are of course not limited to the building

occupants; builders/installers benefit from using less toxic materials during construction and maintenance works, as does society and the planet through the reduced environmental impacts. The building envelope itself can benefit as when well-designed and built natural buildings, in contrast to many common industrial assemblies, don't trap moisture in unwanted ways.

This chapter considers the human health benefits derived from using natural building materials in construction. Of course, not all natural building materials are healthy — asbestos is one obvious example — and all materials carry some level of risk in use, so as well as considering the health benefits derived from natural materials, this chapter considers some of the health risks. The aim here is to provide a current overview of the development, understanding, and application of natural building materials to improve health and well-being. With growing concerns about environmental pollution and poor air quality, this seems to be more important than ever.

Health Benefits

Many of the health and well-being benefits of natural building materials are linked to their passive regulation of indoor air quality, and, in particular, Relative humidity levels, whose links to health and well-being are well known.[6] Readers are likely to have experienced discomfort associated with high RH levels and/or very dry air, which as well as being very uncomfortable can lead to ill health. Relative humidity directly affects our personal comfort since we rely on perspiration

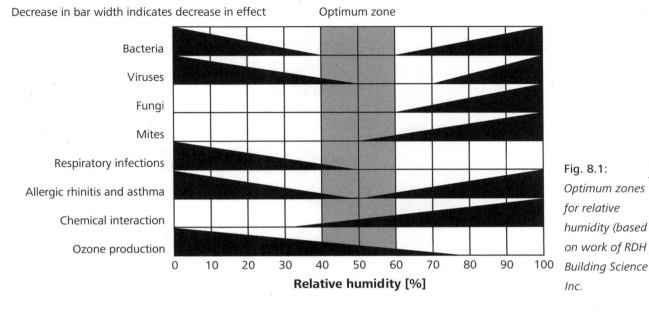

Decrease in bar width indicates decrease in effect   Optimum zone

Bacteria

Viruses

Fungi

Mites

Respiratory infections

Allergic rhinitis and asthma

Chemical interaction

Ozone production

0   10   20   30   40   50   60   70   80   90   100

Relative humidity [%]

Fig. 8.1:

Optimum zones for relative humidity (based on work of RDH Building Science Inc.

(evaporative cooling) to regulate our own body heat. Moisture introduced by normal activity such as cooking and washing can lead to condensation problems, often causing mold growth on internal surfaces and deterioration of the building fabric. As well as directly influencing our own comfort and productivity, extremes in RH are associated with asthma, viral and bacterial activity, and other potentially harmful conditions. As seen in Figure 8.1, maintaining RH levels between 40 percent and 60 percent provides the best conditions for health and well-being; the greater the height of the triangles, shown, the greater the (undesired) activity at low and high humidity.

The benefits of using natural building materials extend to the builder, too. Installation of many human-made materials such as mineral wool and spray foams require personal protection equipment (PPE) to protect the installer from potentially harmful chemicals and particles. By contrast, some natural insulation products such as wool present no hazard to installers.

Moisture Buffering Materials

A number of widely used natural building materials, including clay, wood fiber, sheep's wool, straw, hemp-lime, and timber,

Editor's Note

What follows is a very much condensed discussion of the building science of natural materials, a subject that could easily fill another book.[7] For more general knowledge of building science and related resources, we also recommend RDH consultants in Toronto.[8] There are now good building scientists in most major cities, but the Canadians are to building science as the Japanese are to Sumo wrestling. And be warned: building science just isn't simple, and should never be attempted without adult supervision.

are *hygroscopic*. Hygroscopic materials can adsorb and desorb water vapor from the surrounding air, thus passively and beneficially regulating the indoor environment via moisture buffering. Moisture in the form of water vapor is exchanged with the external environment and the pore structure of the material in response to changing environmental conditions, helping keep RH in the sweet zone of Figure 8.1. Though many industrial materials such as gypsum and lightweight concrete demonstrate similar behavior, the moisture buffering capabilities of many natural materials far exceed those of more conventional industrial materials. Moisture buffering effects are not well appreciated by the construction industry, but as described have significant implications for occupant comfort, health, and well-being, as well as material durability and structural integrity.

A number of small-scale laboratory tests have been devised to measure a material's ability to buffer moisture, including the widely used NordTest Protocol,[9] ISO 24353,[10] and the Japanese Industrial Standard JIS A 1470-1.[11] In these methods, the surface of the test material specimen is exposed to a series of controlled step changes in RH, at constant temperature, for a given time period. The material specimen is placed in a climate chamber, or a vessel containing salt solutions, and weighed periodically to determine the change in mass and thus the change in moisture content. The moisture buffering response of a clay plaster specimen shown in Figure 8.2 below displays the typical sawtooth response to increasing and decreasing RH. When the RH is increased, the specimen mass increases non-linearly, driven by the water vapor pressure, initially at a faster rate, giving rise the ascending sawtooth. Depending on the

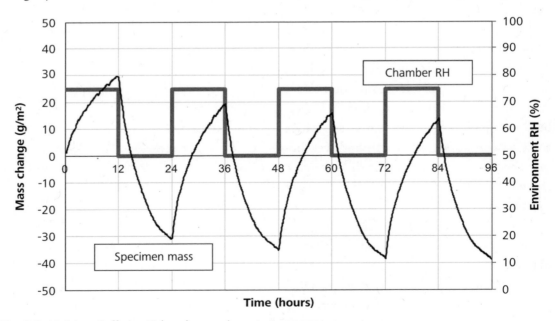

Fig. 8.2: *Moisture Buffering Value clay specimen test response.*

material properties and dimensions, equilibrium may be reached, resulting in no further adsorption or desorption for a given set of conditions. When the RH level reduces, the specimen mass reduces, as shown in Figure 8.2, with the descending non-linear curve which is due to the change in vapor pressure gradient. The moisture buffering value (MBV) is defined by the amount of moisture adsorbed per unit area and percentage RH change.

Table 1 presents MBV values for a range of building materials measured at the University of Bath.[12] A wide range of MBVs is evident, and, importantly, these highlight an order of magnitude difference in MBV between conventional/industrial materials such as gypsum plasterboard and natural fiber materials, such as wheat straw and hemp-lime.

The hygrothermal performance of clay plasters, including the moisture buffering value and thermal resistance, can be improved through material innovation.[13] For example, the inclusion of hemp shiv aggregates into clay plasters improved thermal resistance by up to 80 percent, while moisture adsorption and desorption rates were over 60 percent higher than for the standard test reference plaster.

Studies have confirmed that interior moisture buffering by the building fabric can beneficially affect energy consumption, component durability, thermal comfort, and air quality.[14] Padfield demonstrated the enhanced moisture buffering potential of clay-based materials compared to many conventional materials,[15] and Maskell et al. showed that increasing thickness of clay

Table 1: Moisture buffering values of common and natural building materials

Material	MBV (g/(m²%RH)
Wheat straw	4.62 ± 0.35
Hemp-lime	4.47 ± 0.20
Sheep's wool	2.51 ± 0.13
Hemp fiber	2.01 ± 0.10
Wood wool	2.46 ± 0.11
Modified clay	2.23 ± 0.09
Wood fiberboard	1.80 ± 0.09
Clay	1.54 ± 0.17
Strawboard	1.12 ± 0.07
Gypsum plasterboard	0.40 ± 0.03

plasters beyond a half inch had little further benefit on short- to medium-term moisture buffering capacity.[16]

Though research is helping to establish the benefits of using moisture buffering materials in modern buildings, there is still need for wider recognition of this effect, scientific characterization of materials, and development of design guidance to support better practice. Sophisticated software tools such as WUFI[17] can be used to model the heat and moisture transfer in buildings, but designers and practitioners often require simpler guidance in moisture buffering. In recognition, Rode et al.[18] provided an approximate method of quantifying moisture buffering through the following assumptions: moisture production of a sleeping person can be approximated as 30 grams/hour, so two people sleeping through an eight-hour night would produce 480 grams of moisture. On this basis, the area required to adsorb this moisture can be estimated as 48 square meters for gypsum plaster (or wallboard), but only 10–15 square meters for clay plaster. Clay plasters typically form the finish surface of a wall or ceiling, while gypsum wallboard would normally have a vapor-resistant paint applied, further enhancing the relative performance of the clay finish.

The adsorption and desorption of moisture is *hygrothermal*, in that heat energy is adsorbed and released in tandem with moisture. This makes for a quasi-phase-change behavior of clay and other natural materials; they help cool a hot room and warm a cool one through moisture buffering effects in behavior akin to the effects of thermal mass. The effect is widely known via anecdote, and various researchers have in recent years demonstrated the hygrothermal effects of various natural materials, including rammed earth[19] [20] and bio-based insulation[21] in the laboratory and at full building scale. However, full understanding of the holistic benefits of hygrothermal materials on indoor environment and energy performance is still the subject of ongoing research.[22] At present we know that including hygrothermal materials can significantly improve indoor environment, and contribute to better energy performance.

Table 2: Water vapor resistance factor of some common building materials

Material	μ-value	
	Dry-cup test method (3%–50% RH)	Wet-cup test method (50%–93% RH)
Cellular concrete	7.7	7.1
Gypsum board	8.3	7.3
Concrete	110	150
Cement-lime plaster	19	18
Lime plaster	7.3	6.4

The Breathing Wall Concept: Vapor Permeability and Capillarity

A "breathing" wall, roof, or floor is one that allows moisture vapor to pass through it, a process known as *vapor diffusion*. Breathing walls are of necessity made of vapor-permeable materials. Water vapor will always migrate from areas of higher to lower vapor pressure, be that from inside to outside or the reverse, depending on the climate, time of day, and season. Breathing walls are less susceptible to material degradation because moisture vapor is less likely to be trapped as condensate within the building enclosure — a condition that often leads to mold growth. Together with moisture buffering materials, breathing walls also contribute to the passive regulation of interior RH levels and the healthier conditions shown in Figure 8.1.

The water *vapor resistance factor* (μ-value) is a measure of a material's vapor resistance relative to air. Methods for determining water vapor transmission properties of building materials are described in ISO 12572:2001,[23] and the μ-value for some common building materials are summarized in Table 2.

Liquid water passes through building materials by outright leakage and by capillary action. Capillary-open materials, like traditional lime mortars and plasters, allow water to pass through them, which can assist drying and hinder the buildup of moisture within the enclosure.

The concept of the breathing wall is not to be confused with a lack of airtightness. The term "breathing" is used to connote moisture vapor permeability but not bulk air flow. Airtightness in buildings is typically measured by the number of air changes per hour at a given pressure differential (typically 50 Pa). A certain level of airtightness is essential to control heat loss (or gain) via bulk air movement and create energy-efficient buildings; it is a cornerstone of the Passivhaus certification scheme. It is entirely possible — and generally desirable — to produce an airtight building using vapor-permeable (breathing) walls.

Controlling Volatile Organic Compounds

Volatile organic compounds (VOCs) are a wide variety of natural and artificial organic chemicals that readily evaporate or sublimate from their liquid or solid states then to enter the air. Indoor air pollutants include formaldehyde, benzene, aldehydes, ketones, fragrance compounds, polycyclic aromatic hydrocarbons (PAHs), and flame retardants. Common household sources of VOCs include paints, furniture and fittings, cleaning products, and air fresheners. There is growing literature on the effects of exposure to these various compounds by the World Health Organization. Exposure to the full range of VOCs is found to be common, but there is as yet limited data on the combined effects. For ease of analytical measurements and comparisons, the amounts of VOCs in a given air volume or building are commonly summed and expressed as Total Volatile Organic Compounds (TVOC).

Illness or symptoms that can be caused by poor Indoor Environmental Quality (IEQ) include:

+ headaches
+ eye, nose, or throat irritations
+ dry coughs
+ allergy reactions
+ dry and itching skin
+ nonspecific hypersensitivity
+ insomnia
+ dizziness and nausea
+ difficulty in concentrating and tiredness

To address these issues, two European directives consider VOCs:

+ The Solvent Emission Directive (SED) 1999/13/EC. This directive has been fully implemented since 2007.
+ The Products Directive (PD or DECO) 2004/42/EC. This introduces VOC limits for 2010 that relate to specific products and materials (such as paints) containing VOCs.

Selecting materials and products with low VOC emission rates is one direct and obvious means of minimizing harmful effects, and published standards and product declarations help this process. The VOC emission levels from many natural building materials, as measured by standard test methods, are generally (with the exception of some timber products) low. The VOC emission rates of some natural building materials are provided in Table 3.

Some materials have the capacity to capture VOCs, beneficially removing them from the air. One such material is sheep's wool, a natural insulation also widely used in carpets. The interactions between wool and formaldehyde has been studied since the 1940s, largely based on the use of formaldehyde as a modifier of wool for enhancement of fabrics, and later the effects of sheep dip on the properties of the wool fleece.[24][25][26] In

Table 3: VOC area specific emission rates

	VOC area specific emission rates TVOC	
	µg/m²/h 3 days	µg/m²/h 28 days
Sheep's wool insulation	4	0
Hemp fiber insulation	33	8
Wood fiber insulation	911	160
Hemp-lime insulation	28	19
Lime plaster	25	0
Clay plaster	5	
Uncoated chipboard	42	
Coated chipboard	101	48

2007 Haung et al.[27] studied the use of wool fiber to purify air, concluding that wool was a good adsorber of formaldehyde.

Health Risks

Some natural building materials carry health risks as well as the health benefits listed above. Unlike the benefits, some of the most significant risks to health and well-being are primarily to the builder or installer rather than the building user. Some of the recognized health risks associated with the more widely used natural building materials are outlined in the following section.

Radioactivity

The potential for earthen materials to be sources of potentially harmful levels of radon gas has been a subject of discussion in recent years, especially in Germany. Earthen building materials are formed from the weathering of rocks, many of which contain varying low levels of uranium and thorium, so these elements also exist in earthen buildings.

Though at present there is little evidence to suggest that there is a serious health risk, maintaining good levels of ventilation (as is nearly unavoidable for an earthen wall, floor, or plaster) will further minimize any risk should it exist.

Silica Dust

Silica is present in most natural stones as well as sand and clays. Fine dust, known as respirable crystalline silica (RCS), can be fine enough to be breathed into lungs and cause harm.[28] The health risks from RCS include silicosis, chronic obstructive pulmonary disease (COPD), and lung cancer. These serious health risks are primarily to anyone working regularly with sand, clay, or gravel, who should therefore always follow appropriate industry protective measures such as wearing the necessary PPE.

Handling Lime

Lime in its many forms, including ancient building materials that preceded Portland cement, is popular with natural builders. Lime mortars, plasters, and renders are compatible with a wide range of natural substrates as they provide a more flexible and breathable weather-resistant finish compared with cement-based coatings. It is widely seen as a more suitable binder for bio-aggregates such as hemp, and a more suitable stabilizer for earthen materials.

Care is required when handling all lime materials, as like cement they are highly alkaline and reactive in the presence of water; particular care is required when handling *quicklime*. Workers should always use appropriate PPE to protect against inhalation, and to prevent contact with skin and eyes. Hazards include severe eye irritation or burning, sometimes including permanent damage, irritation, and potential burns to unprotected skin, and irritation if lime dust is inhaled.

Protective Treatments

Occasionally protective chemical treatments and coatings may be applied to exposed surfaces of natural materials as a fire retardant or to improve the durability or wear resistance, such as the use of sodium silicate (water glass) on exposed surfaces of earthen materials. The health and safety guidance of product manufacturers should always be followed.

Concluding Comments

Despite their evident benefits when compared with competing materials and products, natural building materials and solutions have struggled to make much impact in a construction market primarily driven by initial build costs. However, with growing concerns and awareness about health and well-being

Fig. 8.3: *Did you ever notice? When Hollywood wants to portray wealth and opulence on TV or in a movie, they show baronial salons with leather chairs, wood and wool flooring, stone fireplaces, wood paneling, and so on — natural materials. It's what we grew up with and what we love, and now we have the science of why. Or would you rather be surrounded by metal, concrete, and plastic?* Credit: Claytec e.K.

related to buildings, there is a growing impetus to choose natural building materials. At present the links between health and well-being and materials is largely indirect; anecdotal evidence supports improvement in productivity and well-being, but the scientific evidence ties natural materials to the indoor environment (IEQ) rather than directly to human health. Research and development in this area continues, as does the increasing use of natural materials all over the world.

Notes

1. US EPA. 2001. *Healthy Buildings, Healthy People: A Vision for the 21st Century*. United States Environmental Protection Agency, EPA 402-K-01-003.
2. Levin, H. 2007. "IAQ: Indoor Environmental Quality: Current Concerns." Building Ecology. www.buildingecology. net/index_files/publications/IAQIndoor EnvironmentalQualityCurrentConcerns. pdf
3. Yu, C.W.F., and J.T. Kim. 2012. "Low-carbon Housings and Indoor Air Quality." *Indoor and Built Environment*, 21(1), 5–15.
4. Allen, J. G., et al. 2016. "Associations of cognitive function scores with carbon dioxide, ventilation, and volatile organic compound exposures in office workers: A controlled exposure study of green and conventional office environments." *Environmental Health Perspectives* (Online), 124(6), 805.
5. www.eco-see.eu
6. Sterling E. M., et al. 1985. "Criteria for human exposure in occupied buildings." *ASHRAE Transactions*, 91(B), 611–622.
7. And in fact it does: see Jacob Racusin's *Essential Building Science*. www.newsociety. com/Books/E/Essential-Building-Science
8. rdh.com
9. Rode, C., et al. 2005. Moisture Buffering of Building Materials. Number Report BYGDTU R-126. Department of Civil Engineering, Technical University of Denmark.
10. ISO. 2008. Hygrothermal performance of building materials and products: Determination of moisture adsorption/ desorption properties in response to humidity variation. ISO 24353. International Organization for Standardization.
11. JIS. 2008. Determination of water vapor adsorption/desorption properties for building materials: Part1 Response to humidity variation (JIS A 1470-1). Japanese Industrial Standards, Tokyo, Japan.
12. Holcroft, N. A. 2016. *Natural Fibre Insulation Materials for Retrofit Applications*. PhD Thesis, University of Bath, UK.
13. Thomson, A., et al. 2015. Improving hygrothermal properties of clay. 15th International Conference on Non-conventional Materials and Technologies (NOCMAT 2015), University of Manitoba.
14. Janssen, H., and S. Roels. 2009. "Qualitative and quantitative assessment of interior moisture buffering by enclosures." *Energy and Buildings*, 41(4), 382–394.
15. Padfield, T. 1998. The role of absorbent building materials in moderating changes of relative humidity. Department of Structural Engineering and Materials, Lyngby, Technical University of Denmark, 150.

16. Maskell, D., et al. 2016. "Direct measurement of effective moisture buffering penetration depths in clay plasters." *LEHM 2016*, 2016-11-12 - 2016-11-14. Dachverband Lehm e.V., Weimar.

17. WUFI® is an acronym for Wärme Und Feuchte Instationär (translated into English means heat and moisture movement) for software package produced by the Fraunhofer Institute of Building Physics in Germany.

18. Rode, C., et al. 2005. Moisture Buffering of Building Materials. Number Report BYGDTU R-126. Department of Civil Engineering, Technical University of Denmark.

19. Hall, M., and D. Allinson. 2009. "Analysis of the hygrothermal functional properties of stabilised rammed earth materials." *Building and Environment*, 44, 9, 1935–1942, September 2009.

20. Allinson, D., and M. Hall. 2010. "Hygrothermal analysis of a stabilised rammed earth test building in the UK." *Energy and Buildings*, 42, 6, 845–852, June 2010.

21. Lawrence, M., et al. 2013. "Hygrothermal performance of bio-based insulation materials." *Proceedings of the Institution of Civil Engineers: Construction Materials*, 166 (4), 257–263.

22. Lawrence, R., et al. 2016. "In situ assessment of the fabric and energy performance of five conventional and non-conventional wall systems using comparative coheating tests." *Building and Environment*, 109, 68–81.

23. BS EN ISO 12572:2001. Hygrothermal performance of building materials and products: Determination of water vapor transmission properties.

24. Alexander, P., Carter, D., and Johnson, K. 1951. "Formation by formaldehyde of a cross link between lysine and tyrosine residues in wool." *Biochemical Journal* 43, 435–441.

25. Middlebrook, W.R. 1949. "The irreversible combination of formaldehyde with proteins." *Biochemical Journal*, 44, 17–23.

26. Middlebrook W.R., and H. Philip. 1942. "The action of formaldehyde on the cystine disulphide linkages in wool: 1 differing in their reactivity towards formaldehyde." *Biochemical Journal*, 294, 302.

27. Huang, X., et al. 2012. "Experimental study of wool fiber on purification of indoor air." *Textile Research Journal*, 77(12), 946–950.

28. Health & Safety Executive 2013. INDG 463: Control of exposure to silica dust — A guide for employees. HSE, UK.

Buildings
Made of Sky

Snapshots of the New
Carbon Architecture

*Painting above by Nino Gaetani, a
24-year-old man with Down's Syndrome.
At left: Mass timber Metropol Station,
Seville, Spain.* Photo courtesy of Arup

DPR Construction deep green remodel
San Francisco, California. Photo courtesy of DPR Construction

Re-build

If an existing building has decent bones — foundation, structure, and enclosure — it will almost always be better to give it a deep green retrofit than to tear it down and replace it, even with a LEED® platinum or net-zero structure.

Size matters

London, 2008.

The classic cities of the world are lovely in great part because they're not too tall. They are accessible by stair, made of natural materials, and don't choke light and air from people inside or out on the street.

A strong argument can be made, especially now that we can build ten, twelve, or more stories with wood, to keep things that way.

You don't need or want to build tall to get optimal density.

Wood

This is not your father's log cabin — mass timber and the expanding world of wood construction.

Because it makes sense

Wild sculptural architecture is possible, but so is simple, attractive floor space that costs less than steel or concrete structures.

Top: *National Assembly for Wales.* Photo courtesy of Arup © Redshift Photography

Center: *Metropol Station, Seville, Spain.* Photo courtesy of Arup

Bottom: *T3 Building, Minneapolis, Minnesota.* Photo courtesy of Ema Peter, MGA | Michael Green Architecture, DLR Group

But not just any wood . . .

Christ the Light Cathedral, Oakland, California (also on cover).

Raise the standard

If we're going to use more timber in construction, let's work with forests for the long run.

Both of these photos are of certified forestry — under two different standards. The upper photo is the Collins Almanor Forest in Chester, California — the first privately held forest land in the USA to receive FSC (Forest Stewardship Council) certification. Logged six times in 75 years, it is still a vibrant, healthy forest with a broad mix of tree species across a range of age classes, including trees that are hundreds of years old. Photo courtesy of CollinsWood.com

The photo at left is of SFI (Sustainable Forestry Initiative) certified lands also in the Pacific Northwest. In the wake of a major storm, the denuded mountainside has sloughed huge quantities of mud into the stream below, harming water quality and fish — and what soil remains will lose much of its stored carbon. Photo courtesy of David Perry

Girl power

The growing popularity of straw bale and other natural building systems has attracted a wildly disproportionate and welcome number of women into construction.

Photo courtesy of Chris Magwood

Going Uptown

Companies like Ecococon in Europe are building lovely computer-manufactured mass timber structures with straw insulation — net-zero buildings with minimal plastic.

Straw

Hundreds of agricultural by-products and thousands of ways to use them — as finish, insulation, structure, and more.

Straw "Stramit" type panels. Photo courtesy of Ortech

Straw panels.

Wood structure + straw insulation — the Europeans are leading the way. Photo courtesy of Ecococon

Concrete

Back to the future:

The Romans made some of the most iconic concrete buildings ever, such as the Pantheon, using no Portland cement or rebar. Have we forgotten something?

Pantheon in Rome, Italy.

Left: *bioMASON makes bricks using only enzymes from natural bacteria.* Photo courtesy of bioMASON

Right: *Addition at Stanford University by architects Dorman Associates.*

Image credit © SkyHawk Photography / Brian Haux

Lots of ways to make artificial rock

Just a few of the companies to watch: Watershed Media makes compressed earth blocks (above) that perform just like concrete blocks but with much less cement. BioMASON (above left) makes bricks with enzymes but no cement, and Blue Planet (left) makes limestone sand and gravel (carbonate rock) with industrial emissions.

Don't build it, grow it!

We're only just starting to find out what the plant kingdom has to offer.

Meet a fungi

Ecovative has pioneered mushroom insulation that grows until it fills whatever cavity it occupies.

Top left: *Ecovative mushroom insulation in a stud wall cavity.* Photo courtesy of Ecovative

Top inset: *Flexibility of Ecovative mushroom insulation.* Photo courtesy of Ecovative

The amazing grass

Spectacular bamboo architecture is appearing all over the world, taking advantage of bamboo's extraordinary strength. But companies like BamCore have figured out ways to turn those hollow tubes into flat insulated panels for walls, floors, and roofs.

Center right: *A house made with BamCore bamboo panels.* Photo courtesy of Harvey Abernathey and BamCore

Center left: *John Hardy Jewelry showroom, Sibang Kaja, Bali.* Photo courtesy of Darrel DeBoer, Architect

Center inset: *Split bamboo ends.*

Anything but smoke it

Hemp grows like a weed, and provides food, clothing, and now shelter: hempcrete is the natural foam insulation, and can be sprayed or hand-packed.

Bottom Left: *Packing hempcrete in wall formwork.* Photo courtesy of Linda Delair

Bottom right: *Hempcrete in wall.* Photo courtesy of Chris Magwood

Making it

Localizing manufacture and opening up the possibility of zero waste

Chip off the old block

Evolution of the Concrete Masonry Unit, or CMU: start with any fiber, such as straw, rice hulls, shredded plastic, or rubber, and add glue: could be clay, lime, or any of the bioadhesives coming online; compress it for structural strength, or leave it fluffy for insulation — whatever you've got, made on the back of a truck.

Stacking straw blocks (company in stealth mode early 2017).

Any shape you want

Rael-San Fratello Architects are pioneering 3-D printing using everyday materials like sand, salt, and wood.

"Saltygloo" — an igloo made of 3-D printed salt.
Photo courtesy of Rael-San Fratello Architects

"Sound Wall" made of 3-D printed rubber.
Photo courtesy of Rael-San Fratello Architects

Chapter Nine

Size Matters: Can Buildings Be Too Tall?

by Ann V. Edminster

The Height Problem

OUR CULTURE REVERES HEIGHT, whether in people or in buildings. There seems to be an unspoken contest among city developers to build the tallest building — height is an obvious symbol of and proxy for power and economic supremacy. The best views command commensurately higher price tags. But this is a cycle with few winners: tallest today, second-tallest tomorrow — and in certain urban contexts, too far *up* may be a grave mistake when looking through the carbon emissions lens. A more critical look is in order.

I have been chin-deep in the realm of ZE — zero (operating) energy (or emissions; take your pick) — since 2009. Increasingly, practitioners in this realm have focused on cities tackling ZE at the community level as a principal component of their climate action plans. Simultaneously, the phenomenon of living close to San Francisco has faced me with the question: Sure, density is good, but can there be too much of that particular good

thing? More specifically, is further densification (primarily vertical) of an already-dense community moving the carbon needle in the right direction ("good" density)?

My gut tells me that there are limits to how much density is good, and there are a number of indicators this may be true, or true in specific contexts (e.g., when it is achieved via ever-taller buildings and/or in already-dense communities). This chapter represents the beginning of what I hope will be a much longer exploration. Here I've attempted to provide a comprehensive problem statement, identify the relevant indicators, and delineate the research that is needed to answer what may be a quite critical question: Are there limits to how much increases in urban density — especially via vertical growth — will *contribute to* rather than impede achievement of cities' carbon reduction goals? It's quite likely that the answer to this question will differ based on the climate, transit situation, and/or development status of a given locale. For example, the optimum density in a new city designed from scratch — where density can be matched with transit capacity — may be different from the density sweet spot in an existing, already-dense city.

I believe there are a number of potentially sustainable settlement forms: the village, the town, the city, and the metropolitan area (although I also have doubts about the limits of metro areas), all quite loosely defined in my mind, albeit better defined by urbanists. For now, though, while recognizing that there may be different answers as to what constitutes good density for different settlement forms, I am focusing on existing developed urban areas for the simple reason that that's where we have the most people, the most buildings, the most construction activity, and the most potential to cost-effectively reduce carbon emissions.[1] Further, many thought leaders believe that cities represent the best opportunity for impactful carbon-reduction policy leadership.[2] Urban areas are also the only places where I believe we may be in danger of exceeding good density thresholds.

Aspects of the Problem

Height figures prominently in this discussion of density because in existing developed urban areas, increasing density generally translates to increasing height, given the relative scarcity and cost of land. The indicators suggesting the need for density and/or height limits cluster around the following issues:

+ building height as a driver of embodied carbon
+ limits to the mass transit justification for density increase

+ the relationship of height to operating energy
+ livability
+ resiliency

With the exception of livability, all of these arguments have carbon implications.

Ground Zero: Height as a Driver of Embodied Carbon

What started me down this path of inquiry was contemplation of the massive ongoing investment of carbon-intensive resources in new buildings in San Francisco, each surpassing its predecessors in height and skyline domination. Notwithstanding my recent focus on ZE, embodied carbon has preoccupied me for more than 20 years; my master's research[3] involved development of an early whole building LCA method to enable the comparison of embodied energy (as a proxy for carbon), water, and waste resulting from adoption of different building systems.

More recent and sophisticated LCA studies now strongly suggest, if not prove (see Chapter Two), that construction of larger and taller buildings (i.e., those with more stories, both above- and belowground) represents greater carbon expenditure, on a per-square-meter basis, than construction of smaller/shorter buildings. (Choice of structural systems — steel and concrete over wood — is a primary driver of this.) It follows that greater density that relies on bigger/taller buildings leads to higher carbon intensity of construction. Recognizing that the economic forces fueling the development of tall buildings are formidable, I believe we have a compelling obligation to examine whether or not their higher embodied carbon density is justified from a broader carbon — and livability — perspective. Only a very strong case against the increasing height trend will have the faintest hope of arresting it.

Will Transit Catch Up?

When I raise concerns about the carbon impacts of increasing rates of tall-building construction with members of the green building/sustainable development community, invariably the response is, "We need density because that's what supports good mass transit, and we need mass transit to reduce vehicle miles traveled (VMT) and thereby reduce carbon emissions." Sure, that makes sense … where there isn't already sufficient density to support mass transit. But what about, for example, central San Francisco, or Manhattan, where there *is?*

In existing dense urban areas, there are significant obstacles to the creation of new mass transit systems, or even updating existing ones. Those obstacles are logistical, cultural, and financial.

- Logistical: there are a lot of things that need to be worked around — people, streets, buildings, etc.
- Cultural: nobody likes their neighborhood ripped apart, even if what's being created is supposed to (eventually) be wonderful.
- Financial: infrastructure work in developed areas is wicked expensive, because land in urban areas costs a lot … not to mention those pesky logistical and cultural factors.

As a result, in US cities and, perhaps, elsewhere, it takes an extraordinarily long time to add new transit capacity, so transit supply lags behind demand (as manifested by automobile-dominated traffic congestion) by a good many years. That means by the time enhanced transit capacity is in place, unless construction and population growth have come to a halt in the meantime, it already will be behind demand. Again. Meanwhile, as

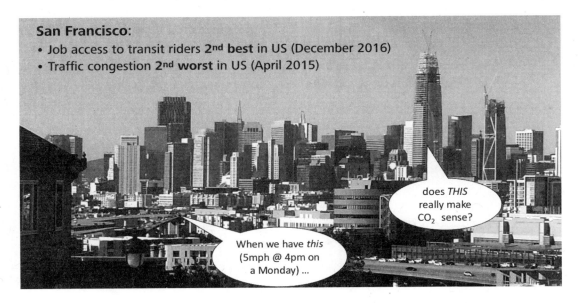

Fig. 9.1.

dense cities further densify by going vertical, automobile traffic and emissions continue to increase, and those emissions are directly attributable to the VMT of all the individuals who now live and/or work in the area whose transit system doesn't adequately serve their needs (and never will, because of the lag described above). It's not a stretch to attribute these emissions to the carbon footprint of the new workplaces and residences built to accommodate those individuals.

In a year or two, New York City's recently opened (December 2016) Second Avenue subway line may provide an interesting case study. The Upper East Side, which it serves, has long been underserved by transit.

In anticipation of the opening of the new subway line, developers have been busily increasing density in this already dense neighborhood. "Even before the first train rolls, the area's housing market has been heating up. New condos have risen on First and Second Avenues ..."[4] This phenomenon is known as induced demand, aka "build it and they will come." These new condos mean more riders on all the transit modes, including the new subway line. It remains to be seen what impact that will have, and for how long (if ever) the new capacity meets demand.

Will the trajectory of induced demand and lag of transit supply ever reach stasis? If not, or even if it that is possible, can further

Affordability: A Brief Detour

Housing affordability is another primary driver of urban densification, or so it is said here in the San Francisco Bay Area. There is undeniable pressure to provide more city housing in our increasingly urbanizing world. Some pundits believe that the unappealing but unavoidable options are to build more housing to maintain affordability or not build more, causing housing costs to skyrocket.[5] Yet this is not an incontrovertible truth; the New York Times reports: "Across the United States, good transit access often leads to higher real estate prices, with home values near rapid transit in Boston, Chicago, Minneapolis-St. Paul, Phoenix, and San Francisco far outpacing other properties during the last recession."[6] So the relationship between mass transit, increasing density in developed urban areas, and affordability is not at all clear-cut.

Continuing to increase density/height without answering this critical question about density/height limits isn't acceptable. We must get that question answered; and then — presuming there *are* limits to height and/or good density — put substantial resources toward figuring out how to densify districts, towns, and suburbs that are too sprawly into places that are more appealingly urban, i.e., dense enough to support transit, with better access to services and amenities, and also containing the more intangible elements of culture that make people want to live there. Affordability should not excuse or delay this inquiry. In fact, the imperative of providing affordable housing in appealing communities only increases its importance.

densification of a city or district that already more than adequately supports mass transit be justified, given the lag of transit supply behind demand, when there is a high carbon cost associated with that densification?

Middle Ground, Perhaps

Low density probably isn't the answer, either. While embodied carbon is of paramount concern, as explained thoroughly elsewhere in this book, operating energy and its associated emissions also remain central concerns for low-carbon architecture. Thus appropriate density and height need to be considered while simultaneously addressing net operating energy/carbon.

Research on net energy use as a function of housing density has shown that new energy-efficient, medium-density housing with photovoltaics can achieve energy use intensities as low as, or lower than, high-density housing, even when transportation energy is factored in.[7] Interestingly, the same study suggests that low-density housing (energy-efficient, with solar electric systems) might achieve even lower energy use intensities than either medium- or high-density housing; however, this is predicated on two conditions that are unlikely to be met, at least in the foreseeable future: "(1) each home maximizes solar collector coverage and (2) the entire personal vehicle fleet is upgraded to be extremely efficient."[8] It also disregards effects related to congestion from reliance on personal vehicles.

While the study authors caution against over-generalizing their findings, they certainly provide sufficient grounds to challenge the presumption that only high-rise housing is to be favored from a carbon perspective.

A more recent study by Adrian Smith + Gordon Gill Architecture (AS+GG) also provides some insight on this issue, though it is noteworthy that AS+GG is a firm which designs very tall buildings. The study modeled communities of 2,000 housing units, using nine different residential building types ranging from detached, single-family suburban homes to a single 215-story building. They analyzed operating energy use, land use, and life cycle carbon impacts, including embodied carbon of the buildings themselves as well as associated infrastructure. The study showed that the 4-story courtyard building was the most energy-efficient of all the prototypes, and also represented the lowest life cycle carbon intensity.[9]

From a land use perspective, they concluded that the 58- and 34-story models were the best performers, followed by the 16-story and then the 4-story courtyard. However, this analysis was based on unrealistic assumptions — e.g., a 20-year building life cycle, which is far less than the average building life span of 120 years.[10] Also, the only value they assumed for the land unoccupied by buildings was for photovoltaic (PV) electricity generation, and they did not account for the PV generation potential of the buildings themselves. These assumptions seriously compromise the credibility of their land use conclusions.

Another component to the argument for moderately tall buildings over very tall ones is the form factor. The footprints of very tall

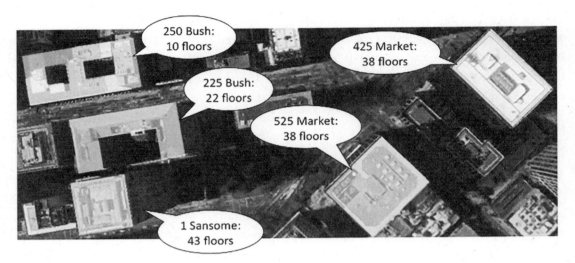

Fig. 9.2: *Downtown San Francisco buildings. 225 and 250 Bush Street are classic examples of building whose height allows freedom in plan form to admit light and fresh air to the interiors while also self-shading some of the windows. By contrast, the block-form towers exemplify the way wind forces in taller building drive the use of more compact plan shapes, which admit less light to the interior and provide no self-shading. (The more extensive ground-plane shading from the taller building is also evident from Google Earth views.)* Credit: Google Earth, Hal Bohner and Ann Edminster

building are typically convex shapes — all their corners point outwards. This type of footprint drives a number of things: a shell dominated by glazing, with consequent thermal, energy, and comfort challenges; and thus increased reliance on air conditioning and artificial lighting over natural ventilation and daylight.

By contrast, moderately tall buildings are structurally better capable of assuming open forms such as C- and H-shapes, which allow windows to be better distributed relative to floor area for solar control, daylight, and natural ventilation. These building forms can also be designed to self-shade, to some degree.

A study of New York City benchmarking and disclosure data found that "older buildings ... often consume less energy and are less carbon intensive ... than newer 'high performance' buildings [due to having] quite simple heating systems, with fewer fans and pumps than are found in newer buildings."[11] A deeper dive into these buildings might reveal that form also plays a role in the lower energy intensity of the older buildings. For example, lighting consumes more energy than any other electrical end use in New York City's office buildings,[12] and form certainly plays a significant role in lighting energy use:

> Many of the City's older office buildings were designed to utilize daylight, as they were built in electric lighting's infancy. Block sizes and orientation have generally resulted in prewar building floor plates in which daylight

reaches a good portion of the floor area — in fact, New York City's first comprehensive zoning ordinance was enacted in 1916 in part to preserve access to daylight. Apparently, the 1916 zoning changes were accepted by building owners in part because they understood that daylit offices could command higher rent.[13]

It would be illuminating (pun intended) to analyze the extent to which New York's office lighting *needs* (not demand, as demand can be changed with high-efficacy lighting retrofits) are based on floor plate geometry, and how that correlates to height.

Another argument in favor of moderate height, as noted in Chapter Four, is that all-timber buildings (i.e., the structural alternative lowest in embodied carbon) are likely to be most cost competitive in the range of 6 to 12 stories. At this height, wind comfort, fire resistance, and structural strength can be achieved with relatively slender cross-laminated timber panels and simple connection details. At greater heights, structural complexity increases, along with the need for steel and/or concrete.

A final aspect to the form aspect of carbon intensity is the extent to which a building's energy performance is dominated by internal loads (people, lighting, equipment, etc.) versus its skin, or enclosure. Small buildings are generally skin-dominated while tall buildings are generally internal load-dominated. A design driven by balancing internal loads and enclosure loads might also point to a middle ground between small and tall buildings.

Livability

Access to daylight and fresh air are important not only to reducing energy intensity, but also to livability. As human beings, we evolved in the presence of both fresh air and daylight — we need them for our physical and mental well-being. This is important both in homes and in work environments, where many of us spend the majority of our daylight hours. As mentioned above, moderately tall buildings offer improved opportunities to incorporate both natural ventilation and daylight into living and working spaces than do very tall buildings.

Another aspect of livability favoring shorter over taller is the effect of buildings on street life. Anyone who has spent much time in a high-rise downtown area can speak to the unpleasant wind tunnel effects, excessive shading, and dehumanizing environment created by super-tall buildings and their over-wide, imposing facades.

Equally important are the psychological and sociological impacts of excessively tall buildings. In *A Pattern Language*, Christopher Alexander and his collaborators speak to this eloquently:

> There is abundant evidence to show that high buildings ... can actually damage people's minds and feelings ... There are two separate bodies of evidence for this. One shows the effect of high-rise housing on the mental and social well-being of families. The other shows the effect of large buildings, and high buildings, on the human relations in offices and workplaces.[14]

The authors then posit an explanation:

> High-rise living takes people away from the ground, and away from the casual, every-day society that occurs on the sidewalks and streets ... The decision to go out for some public life becomes formal and awkward; and unless there is some specific task which brings people out in the world, the tendency is to stay home, alone. The forced isolation then causes individual breakdowns.[15]

These views may sound somewhat melodramatic, but the authors offer numerous studies related to both housing and workplaces to support their view that, "throughout the city, even at its densest points, there are strong human reasons to subject all buildings to height restrictions."

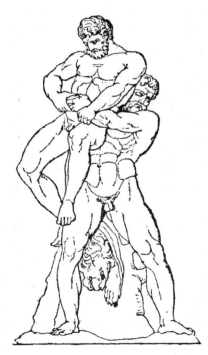

Fig. 9.3: *The notion that we belong close to the Earth is nothing new. Myth holds that Antaeus, Gaia's half-giant son, reigned supreme among Greek wrestlers so long as he remained in contact with his mother, the Earth. Unfortunately, Heracles got wise and defeated Antaeus by holding him aloft and crushing him to death.* Credit: Wikipedia

> *Man is an outdoor animal. He toils at desks and talks of ledgers and parlors and art galleries but the endurance that brought him these was developed by rude ancestors, whose claim to kinship he would scorn and whose vitality he has inherited and squandered. He is what he is by reason of countless ages of direct contact with nature.*
>
> — James H. McBride, MD (Journal of the American Medical Association, 1902)

Resiliency

How resilient is a tall building? The taller it is, the more reliant on mechanized vertical transportation, comfort, and life safety systems. In this era of escalating extreme weather events and accompanying power outages, every building should be scrutinized for its vulnerability to those occurrences. In a tall building, the likelihood of occupants becoming trapped — however temporarily — in unpleasant and/or dangerous conditions intuitively can be seen to increase as a function of distance from the ground. This risk is greatest for those with mobility limitations, for whom navigating stairs may be difficult or impossible. This should be of grave concern, for example, in housing for the elderly.

From a resiliency perspective, there is also evidence suggesting that something between tall and short may be optimum. New York's Urban Green Council advises, "Not all buildings hold their temperature equally well without power. The brick walls of row houses and low- and high-rise apartments hold some heat, and newer buildings tend to be better insulated. On the other hand, single-family houses are exposed on four sides, and all-glass buildings [typical for tall buildings] lose heat through their windows in winter and gain it in summer."[16]

Conclusions

Much as I trust my own instincts (don't we all?), instinct isn't enough to resolve the questions I've raised here; we need quite a lot of research to determine what, if any, limits exist to climate-neutral urban density.

1. What level of density will support mass transit — and other low-carbon transportation alternatives — that is good enough (in terms of proximity, frequency, service quality, affordability, etc.) to supplant personal automobile use?

2. Is there a height at which embodied carbon (per functional unit, generally per square foot) starts to increase beyond values consistent with cities' carbon reduction goals?

3. What is the relationship between height and density? Should one or the other — or both — be the focus of policy attention?

4. Are there variables other than building height (e.g., size of footprint) that are also significant drivers of embodied carbon?

5. How long *does* transit infrastructure development lag behind supply? And does supply ever meet demand?

6. What is the carbon load of traffic congestion increases resulting from densification that occurs in the absence of adequate mass transit capacity to serve the population increase?

7. What is the net operating energy (factoring in both energy use and solar energy production) sweet spot for building height/density? To what extent does that sweet spot change based on climate, for example, and other variables?

8. To what extent are lighting and other energy loads influenced by building form?

9. To what extent is building form dictated by height?

10. What is the relationship of floor plate shape to access to daylight, views, and fresh air?

11. To what extent do tall buildings drive wind tunnel effects, shading of sidewalks, and other urban open spaces?

12. How important is building footprint size to the quality of pedestrian experience, and the quality of street life in general?

13. Is there current research looking at the relationship between building height and well-being?

14. Resiliency is of central concern to cities; is there a solid correlation between resiliency risks and building height? Do development policies that are blind to building height adequately safeguard building resiliency?

Notwithstanding the abundance of unanswered questions — or perhaps because of them — I believe there is ample evidence to justify a serious inquiry into this complex issue. In the meantime, a number of carbon and livability indicators suggest that cities would do well to limit building height to 10 or 12 stories.

Editor's Endnote

Ann's arguments against excessively tall buildings and cities are compelling, or at least, as she states, should be seen as the beginning of a much-needed debate. The subject may seem less directly tied to the theme of this book, but in fact is an important part of visualizing and moving toward a climate-friendly architecture.

Size matters. And, having made the argument for restraining height, we need also at least mention building volume. In particular, I couldn't end this chapter without noting the trend in affluent communities everywhere to build excessively large single-family houses. Unlike height, this is a subject that has received plenty of discussion and heated debate that we won't try to replicate here.[17] Let it suffice to opine: excessively large homes are a climate disaster, arguably somewhat less so when they achieve some sort of green or energy-efficient certification. The increasing trend of second homes — many of which remain unoccupied for a large part of the year — is another concern. Of the spate of tall residential towers sprouting in New York City, *Architectural Record* observes, "One complaint is that the apartments' uberwealthy investors will rarely occupy them, leaving the buildings — most of which are concentrated around the southern end of Central Park — empty and lifeless."[18] In this case, the climate offense is compound: these buildings are extravagantly tall, large, *and* frequently vacant.

All of this begs the question of who gets to define "excessive"; excessive in one town or culture is just a little bungalow in another. Let that debate go on with our simple input: small is beautiful. And, to the libertarians demanding to be left the hell alone to build what they want, we add the simple dark companion to that view, so well-known to building officials everywhere and glossed over by libertarians: *everyone* wants to be left to do whatever the hell they want — but also want government to make sure their neighbors don't do something they might not like. Governance would be easy, to paraphrase Yogi Berra, if it weren't so hard.

Still, there may be a bright future for the "McMansions" currently sprouting like mushrooms all over the North American (and other) landscapes. With some adjustment to current planning rules, they could easily be turned into cooperative housing for multiple families, reducing in several ways their carbon footprints. Other emerging trends are also promising: the tiny house movement (which generally calls for minor adjustment to building codes), and the auxiliary dwelling unit movement that would foster infill of suburbs (and typically requires minor adjustment to common planning rules born in the postwar era).

Building large, like building tall, satisfies certain urges, and makes money for a select few. But what is the point of having a nice house, as Henry David Thoreau once said, if you don't have a decent planet to put it on? We heartily salute and support all efforts to rein in building practices that benefit a very few but penalize everyone. On a crowded planet, that's not in anyone's best interest.

Notes

1. "Urban areas account for up to two-thirds of the potential to cost-effectively reduce global carbon emissions." International Energy Agency (IEA), June 1, 2016. www.iea.org/newsroom/news/2016/june/etp2016-cities-are-in-the-frontline-for-cutting-carbon-emissions.html

2 "Cities Must Take the Lead in the Transition to a Low-Carbon Energy Sector." IEA 2016. Wired Mag, December 2016, "Nations Be Damned, the World's Cities Can Take a Big Bite Out of Emissions."

3. Edminster, *Investigation of Impacts: Straw Bale Construction*, Master's thesis, University of California Berkeley, 1995.

4. *New York Times.* December 30, 2016. "Second Avenue Subway's Arrival Brings Fear That Rents Will Soar."

5. www.citylab.com/equity/2016/09/the-difficulties-of-density/ 499571/

6. *New York Times.* "Second Avenue Subway."

7. O'Brien, W., et al. "The Relationship Between Net Energy Use and the Urban Density of Solar Buildings." *Environment and Planning B: Planning and Design 2010,* 37, 1002–1021. November 2010, 15, Figure 5.

8. O'Brien, 8.

9. Drew, C., et al. "The Environmental Impact of Tall vs Small: A Comparative Study." *International Journal of High-Rise Buildings,* June 2015.

10. Donnelly, B. "The Life Expectancy of Buildings." 9/6/2015. brandondonnelly.com/post/128489870433/the-life-expectancy-of-buildings

11. Hinge, Adam, et al. 2013. "Building Efficiency Policies in World Leading Cities: What Are the Impacts?" *ECEEE Summer Study Proceedings,* 778.

12. Energy Efficiency Potential Study for Consolidated Edison Company of New York, Inc.; Volume 2: Electric Potential Report, Global Energy Partners, LLC , Walnut Creek, March 2010.

13. Building Energy Exchange. 2012. "Let There be Daylight: Retrofitting Daylight Controls in NYC Office Buildings." be-exchange.org/resources/project/31

14. Alexander, C., et al. 1977. *A Pattern Language.* New York: Oxford University Press, 115–119.

15. Ibid, 169–472.

16. Urban Green Council. 2014. "Baby It's Cold Inside: Can NYC Buildings Protect Us from Severe Weather During Blackouts?" urbangreencouncil.force.com/BabyItsColdInside

17. Noteworthy in this context is Ann Edminster's work in the early 2000s to develop a size adjustment mechanism in the US Green Building Council's LEED (Leadership in Energy and Environmental Design) for Homes program to account for the environmental impacts of homes as a function of their size.

18. Gonchar, Joann. "Stacking the Deck." 2017. *Architectural Record.* May. www.architecturalrecord.com/articles/12677-leonard-street-by-herzog-de-meuron

Chapter Ten

Technology and Localization: Trends at Play

Will this "fourth industrial revolution" lead to an acceleration of the extractive, "linear" economy of today, or will it enable the transition towards a society in which value creation is increasingly decoupled from finite resource consumption? Intelligent assets are a key building block of a system capable of ushering in a new era of growth and development, increasingly decoupled from resource constraints.[1]

TECHNOLOGICAL INNOVATION HAS been happening for centuries, weaving through and underlying everything in this book. From the recent manufacturing developments that make mass timber structures and straw panels possible, to the molecular-level inventions of new cements and artificial rock, to the web-enabled data crunching that makes life cycle analysis possible, we are ever inventing new ways to do things — in the mantra of Silicon Valley: better, faster, cheaper.

But there's more, so very much more. Many powerful trends are at work altering and disrupting all of human life and industry, even if only just beginning on the built environment. From mud huts and grass shacks to transit centers and office towers, we're keeping and reinventing every form of shelter we've ever had, and we're reimagining "building" from the molecular to the global level. We're reinventing how we think about it (design and engineering) how we gather and assemble atoms into bricks, buildings, and cities (manufacturing and construction) and how we manage and track the whole messy business (industry, trade, and governance).

And we're just getting warmed up:

The speed of the current breakthroughs has no historical precedent. When compared with previous industrial revolutions, the Fourth [Industrial Revolution] is evolving at an exponential rather than a linear pace. Moreover, it is disrupting almost every industry in every country. And the breadth and depth of these changes herald the transformation of entire systems of production, management and governance.[2]

When I was in engineering school in the late 1970s, I worked evenings and weekends as a roofer. As the new guy, my first job was humping — picking up the bales of shakes or shingles that had been dumped on the ground by a supply truck, and then carrying them one by one up a ladder to the roof. By the end of summer, I had legs of steel and a nice bit of cash. The following summer, I graduated up to roofer, and became very facile at grabbing a handful of galvanized 10 penny roofing nails with my left hand, arranging them, and feeding them one by one to my right, which holding the roofing hammer then set them and drove them home. Set, drive, set, drive, set, drive in ever quickening pace; we got paid by the area covered, not by the hour, and were very motivated to get good at it. By the time I graduated from school, both jobs had been made substantially easier by technology: a gas-powered portable conveyor belt that ran to the roof, and electric nail guns. It was great! The march of technological progress that had begun with Watt's coal-powered steam pump two hundred years earlier looked like nothing but good news.

Forty years later, things have sped up a bit. Trends such as nanotechnology, biotechnology, robotics, artificial intelligence, and 3-D printing (also called additive manufacturing) are evolving and comingling at a dizzying pace in ways that are only just starting to affect the way we build, what we build

with, or how we even conceive "building." All of those trends are themselves superpowered by the acceleration in the metatechnology of massive data storage and ever faster connectivity, aka "the Cloud." As biologist Janine Benyus put it:

> *The Internet of Things and circular economy practices are mutually reinforcing — bundled together they present immense opportunities — for business and society at large — leading to systems that are resilient, decentralized, self-repairing and scalable without experiencing complexity problems . . . and present enormous economic opportunities to plug leaks and make use of materials previously considered to be waste.*[3]

More than a few large books could be written just about technology and architecture — indeed, many already have — and they would of course be obsolete in a matter of a few years or weeks or minutes. Still, in talking about building with carbon, we should at least take a brief look around at the landscape in early 2017 to assess how these trends may affect how we build. Your forgiveness is asked in advance for the inevitable omissions, sweeping generalizations, and wild conjecture, but, hey, we had to give it a shot.

Nanotechnology

To date, nanotech is mostly a force in the medical and electronics industries, but there have been a few visible forays into buildings. Specialty coatings that resist decay or absorb pollutants are now available, as are the aerogel family of thin insulation systems. Metals, concrete, and novel ceramics have been strengthened in laboratories and/or improved vs. carbon footprint, but have not yet shown up much on jobsites. As is typical for any new family of products, nanotech materials are expensive, and show no clear trend as a carbon sequestering or saving strategy. Further, their very existence raises fear, well founded or not we don't yet know, about releasing nanoFrankensteins into the environment or into our bodies. Much promise, to be sure, but also much risk we haven't yet fully assessed. Regardless, we will surely be seeing nanotech innovation entering the construction materials marketplace on many fronts. And it's safe to predict that regulation and consumer/environmental protections will, as always, lag behind the development and use of new materials.

Biotechnology and Biomimicry

Nature to be controlled must be obeyed.
— Sir Francis Bacon

Similarly, biotech is a fast-growing field confined mostly to medicine and food — at first glance. However, as Janine Benyus implies in the quote above, the emergence of systems thinking founded in ecology is starting to alter the way we think about the process of building. It's not your father's brick, nor is it your father's roofing tile, window, or construction site.

Let's step back for a moment. A *building* is a device for protecting human beings, and as such is comparable to, among many examples, the shell of a crab, the feathers of a bird, or the bark of a tree. Though we tend to think of buildings as static objects, they have never been anything of the sort, and are constantly subject to weathering, decay, and the many abuses of nature that they protect us from. Furthermore, they are dynamic centers of energy, material, water, and other flows, including we human users coming and going through them; buildings are the stuff of dreams and myths. As Winston Churchill said, "We shape our buildings; thereafter they shape us." We have tried cranking out homes and buildings like toasters off of a factory production line, and it actually hasn't worked out very well. What if we thought of buildings like a crab thinks about its shell (yes, I know crabs probably don't think, but bear with me here), or a bird thinks about its feathers, or a tree its bark. Can we imagine our constructed shelter as a living organism around us? Can we "grow" buildings, or, anyway, make them without so much fuss, noise, and stink?

Nature offers us plenty of examples for study. You want concrete? The giant termite mounds of Africa, cemented together entirely by the enzymes in termite spit ("Termite spit concrete" — there's a marketing trope!) are hard, strong durable structures. The coral reefs of the world have been built up, bit by bit, by tiny organisms that pull calcium and carbon from seawater to make their fabulous carbonate homes. You want fabric? Spiders spin silk with remarkable properties far superior to anything humans make. One researcher speculated that if he had enough spider silk to make half-inch-thick rope, he could build a net that could catch and hold a 747 in flight. And, in contrast to our typical "heat, beat and treat" methods of material manufacture, all of these natural materials are made without fossil fuels, added heat or pressure, or toxic wastes. The question is, then, can we get the clue? Maybe we can't actually "grow" buildings, but maybe we can create them without all the flames, noise, and stink of modern industry. Maybe we can get a little more elegant in the way we go about making shelter, if only for our own good.

Actually, we've already started. More under the rubric of *biomimicry* (study and imitate how Nature solves problems, and engage her partnership) as opposed to *biotechnology* (modify life forms, typically microbial, to achieve specific purposes), many firms are appearing with promising developments.

Ecovative, for example, started with the study of fungi — mushrooms — and developed a way to mix fungi with water and cellulose pulp, a very common waste product, to get a "spray-on" padding and insulation that naturally grows into the shape of whatever cavity it occupies. They started with product packaging, replacing the polystyrene blocks and peanuts that we all grew up with — a terrific development all by itself — and then moved to buildings, where they are just starting to provide wall insulation. Soon enough, you'll be able to gather whatever paper, cardboard, sawdust, and other cellulose as may be handy, mix it up with some water and magic juice, and spray it into your building for a very effective and inexpensive alternative to the problematic plastic spray foams so common today.

Another biomimetic product comes from BioMASON, who at present offer paving and face bricks made without baking or cement. Making use of naturally occurring enzymes from bacteria, BioMASON produces useful, lovely products with a fraction of the carbon footprint of their competitors. (See photos of both Ecovative and BioMASON in center color section.)

Localization: The Convergence of Social and Technological Trends

As quoted earlier, Janine Benyus described natural systems as *resilient, decentralized,* and *self-repairing.* By implication, and as articulated in her writings, these are proven qualities for systems that work — systems that don't inadvertently end up fouling their own nests or altering the climate. That is, natural systems that by definition are sustainable because they're still here after all these years and millennia.

We're just starting to realize that many technological trends pull toward localization (or decentralization, in Janine's terms), and thus join with some purely social trends as are increasingly visible around us (more on that in a moment), and arc toward more resilient, decentralized, self-repairing, and scalable life — and building.

Robotics, Artificial Intelligence, and 3-D Printing

The convergence of new technologies will dramatically change how we make things, what we make, and where we make them. Trends in energy production, agriculture, politics, and the Internet will accelerate these changes, retarding, if not entirely reversing, globalization. Over the next decade or two, these trends will result in the localization of manufacturing, services, energy, and food production. . . . The combination of robotics, artificial intelligence, and three-dimensional (3-D) printing is rapidly changing how we produce goods.[4]

Just about every part of the economy — farming, energy, building, and more — is already changing and, at least in some ways, localizing. There are, for example, already enormous prototype 3-D printers that can produce a full-size house of concrete or stabilized earth, and the mills that produce those mass timber structures are full of smart machinery to cut, assemble, glue, and clamp wood into the shapes we desire. Straw, so abundant and promising as an insulating material, is light and fluffy and royally difficult to move around; but no problem! It's increasingly easy to create mobile manufacturing facilities that go to the fields as the crops are available. Most of the workers on jobsites have smartphones they can operate by voice command, and which will talk back to them. Tech is here, more and more, faster and faster.

That, and all the localizing effects of innovation, are good news for the climate. It means we can ease back on shipping big, heavy stuff around and start to make our clothes, medicine, bricks, and food much closer to where we buy and use them. We are increasingly able to gather, sort, and use dispersed resources like straw or the bottles, paper, and plastic at recycling plants, enabled by technological advances and the best resource of all,

human ingenuity. Localized manufacturing also produces less waste and pollution, and uses much less energy and water.

For better or worse, it also doesn't need much labor.

Forty years ago, we roofers celebrated the appearance of conveyor belts and nail guns, and because the economy was booming didn't really notice that they eliminated jobs. But our boss noticed: he could now get a roof covered with two thirds the cost and a lot less bellyaching from the now smaller crew. Machines made him money.

This aspect of localization is shaping up to be bad news for manual, skilled, and even highly skilled labor, as on the factory floor and construction site, robots large and small, smart and dumb, stationary and mobile are taking over because they're cheaper and more reliable than human workers. This was underscored for me when, 40 years after my fabulous roofing career, I was a greying engineer serving as advisor to a materials startup company near San Francisco and taken on a tour of the new manufacturing plant with the other advisors. At one point, the plant foreman pointed out a new smart robotic machine, which he proudly described as "paying for itself in three years because of all the workers it replaced." Everyone sagely nodded their heads, pleased, and confident that the investors also would be gratified. Machines, we understood, work 24/7 without complaint, pregnancy, illness, or any vestige of a personal life, while workers are a lot of hassle and expense. Who needs 'em?

In the USA, manufacturing is in fact returning to our shores, but it's just not bringing many jobs. The T.X. Hammes paper I quoted from at the beginning of this section was directed to a military and international security audience whose job is to track global trends; it is of great interest to them to note what technological trends are, and how they affect people. If a Ford or Carrier or John Deere factory returns to Ohio or Kentucky but only employs a small handful of managers and highly trained engineers, leaving the townies standing hungry at the fence, that can start to affect local and international security. It's no great stretch to say that this trend is part of what fueled Brexit and the election of Donald Trump.

The dramatic rise in mechanization raises the question: "If all our jobs are taken away unless we happen to be Bill Gates or Mark Zuckerberg, what will we do? How will we buy anything?" Will we, as many suggest, all receive a guaranteed income from the government, freeing us (if that is the word) from the need to work? That may or may not prove to be part of our future. But we do want to take this discussion to an emerging and somewhat countering trend, less visible, less technological, and less distinct, but potentially just as impactful. That trend is: people unplugging. Technology is driving localization, but is also driving us crazy.

More than a few people, though affluent enough to access the best and most of modern life, choose in a sense to work more, not less. More than a few think that they'd prefer, for example, to grow at least some of their own food, get around on a bicycle, read physical books and magazines, or build their own homes with their own hard work and friendly neighbors. Call them the modern descendants of the so-called counterculture of the 1960s, or of the two-hundred-year-old Luddite movement,[5] they are unlike the stereotypical 60s "hippie in the woods" for two important reasons. One, they live among us, not removed to the deep woods; they feed the goats in the morning and then bike to work at the hardware store in Sydney, or the startup in Madrid, or to teach linguistics at Oxford. Second, their rejection of technology is selective, not blanket; they work the garden all day, then write code at night; they bike rather than drive; they spend an evening playing guitar with friends rather than watch Netflix. Sometimes, or sometimes not. In other words, they are you and me as we rediscover the pleasure of unplugging, making, and physically connecting. Sometimes the digital world is just too much or off target, and we increasingly return to analog: play a record rather than pull out an iPod, or read a real paper book, or in dozens of different ways unplug from the digital and connect with "meatspace." We are, after all, social primates, as is underscored by the dark fact that our worst form of punishment is solitary confinement, or by the general sense that a good time is a party — having fun with other people.

But what, you might ask, does this have to do with building materials? How might the hypertechnological trends of the day converge with un- or even antitechnological social trends to alter the way we build, and the things we build with?

For me, the beginning of an answer appeared in Haiti, and it was this: people are smart and can do almost anything, including building well, if given the tools and knowledge they need. It's unfortunately also true that people are smart but will do dumb and injurious things if they don't have the right tools and knowledge. We sometimes build spectacular architecture with mere mud, and sometimes build crap with stainless steel and imported marble.

Following the devastating 2010 earthquake in Haiti, I traveled to Port-au-Prince under the auspices of the Ecological Building Network (EBNet, the non-profit group that sponsored this book) with the hopes of assisting reconstruction. I saw many jaw-dropping things in a mere week there, but a few stand out in memory. Despite being a lovely, intelligent, and dignified people, the Haitians can't get an even break; they are forever subject to earthquakes, hurricanes, disease, an entrenched and corrupt plutocratic government, and especially a near-total loss of topsoil (I didn't know until I saw it with my own eyes how devastating it is to lose your soil). And they get a lot of technology they don't need. Following the earthquake, there appeared literally hundreds of well-meaning charities and NGOs, each loaded with some version of "house-in-a-box" one-size-fits-all shelter systems, almost all of which were for one reason or another not very helpful. We (EBNet) entered a housing competition sponsored by the Clinton Foundation to develop new solutions for Haiti, and were among the winners for developing an answer to the question (as we phrased it); "How could Haiti rebuild if it could only use what it already has? What if there were no imports?"

In that particular context, we took note of the fact that they had a lot of two things: concrete rubble and hungry, eager-to-work people. Put that together with some simple, hand-powered machines for crushing the rubble and then making concrete blocks with the results, add a bit of fancy seismic engineering (seismic isolation joints made with reused soda bottles!), and you get some steady employment plus cheap, safe housing designed to meet Haitian cultural norms but not fall apart when the ground shakes.

Again: people everywhere are smart, and can do almost anything if given the tools and knowledge they need. Throughout history, human beings have used what was at hand along with the accumulated knowledge of generations of builders to make shelter. They learned by trial and error, refining ways of building highly attuned to place and local resources (with the general exception of earthquake safety; for various reasons, few cultures in history adapted effectively to seismic risk). Today, and for the past century or so, we haven't had to do any of that; we can import windows from Germany, granite countertops from Italy, and cheap steel beams and electronics from China, all to be installed by a hired contractor who, more likely than not, is far too busy to really understand how all these things go together.[6] We built and continue to build energy-hogging glass office towers in Miami and in Montreal. Now we find ourselves surrounded by, dependent on, and immersed in systems we don't understand, so it comes as no surprise that some of us, some times, want to understand more about the systems we live with, or want to just unplug. We would rather build or fix up a home or school with our own sweat, injured hands, neighbors, and mistakes than simply write a check to a stranger. We even have a term and growing market niche for this: DIYs — Do It Yourselfers. When you do it yourself by choice, you join the millions of people who, throughout the world and history, do so of necessity because they're poor and that's their only option.

Either way, there is a limit; building well takes a lot of energy and a lot of smarts, no way around it. If your need is as simple as getting out of the rain for a night, all you need is a tent or a cave or a highway overpass. But as you start to add all the goodies that many of us take for granted, like indoor plumbing, safe and steady electrical supply, watertight roofs, warmth in winter, privacy, security, and so on, then things start getting quite a bit more complex. Almost anybody can build a small adobe hut or repaint a bedroom, and people do all the time, but few have the skills and resources to put together a modern, urban, multistory, multiuse complex.

Web-enabled knowledge is rapidly making it possible for any of us to do things we never really could before, and to build better, with more different products and materials, than we ever really could before. There's a YouTube video at hand now for just about anything you can imagine, and access to all the world's knowledge is a just few clicks away. But there's a limit: knowledge isn't wisdom, and that YouTube video, though very useful indeed, is no substitute for an experienced teacher. Really useful skills come from working with and learning from others, from those with more experience. (The glaring exception that proves the rule is the ever turbulent and disruptive climate of Silicon Valley and its counterparts around the world where developing unprecedented ideas and products is the everyday norm.) To do it yourself means to do it with others.

Building well will depend on access to knowledge, and access to each other and the accumulated experience of building. In fact, access to knowledge and energy has always been key. Look at history: the truly signature innovations changed one or the other. The Gutenberg press, radio, the electrical grid, telephone, and TV, digital computing, photovoltaic power, and the internet. In that context, it's hard to say which of these many technological currents will affect us or the construction industry most. But our screamingly increased access to more and more knowledge has not led to what we often really want: connection. We are social primates, and a like on Facebook is just not the same as a hug. Which is why I suspect that "building" will increasingly be, for some, at times, like growing food or making music: a way — an excuse, as if we needed one — to work and play and be together. It is a vehicle for building community.

Our ancestors relied on the accumulated wisdom of their elders, passed down verbally from generation to generation. Today, and increasingly, anybody can have access to the accumulated experience of the whole world — but it only works if that knowledge is well vetted and curated. With the right voice speaking through your smartphone, you and your neighbors and your pet robots — or a small or a giant construction company — can accomplish miracles. We still rely on the accumulated wisdom of our elders — and you would be unwise to attempt building without their counsel — but that wisdom is increasingly available on a screen in your hand.

*Not all that can be counted counts
and not all that counts can be counted.*

— various attributions

Notes

1. Ellen MacArthur Foundation. 2016. *Intelligent Assets: Unlocking the Circular Economy Potential.*

2. Klaus Schwab. 2016. *The Fourth Industrial Revolution: What It Means; How to Respond.* World Economic Forum.

3. *Intelligent Assets.*

4. T.X. Hammes. 2016. *Will Technological Convergence Reverse Globalization?* Center for Strategic Research, Institute for National Strategic Studies at the National Defense University.

5. It bears remembering here that the Luddites were not opposed to technological progress per se. They burned and destroyed mechanical looms that had replaced jobs and were stealing livelihoods. They were terrified, as we probably should be now, that the machines would eviscerate the well-being of individuals and communities.

6. This is said with a bow to contractors everywhere, who face the herculean task every day of knowing everything about everything, organizing herds of herds of cats, and communicating clearly to clients what they can and cannot expect. Yikes, I couldn't do your job!

Chapter Eleven

Action Plan: Places to Intervene in a System

W<small>E HAVE SKETCHED OUT</small> the new carbon architecture, but how can we hasten its arrival? How do you move a system as massive and complex as the global construction market?

The short answer is pretty simple: carbon taxes and an informed populace. But let's be a little more methodical and detailed in laying out tactics and strategies. To do so, we'll loosely follow the famous list by the late Donella Meadows in *Places to Intervene in a System*[1] as a framework:

Places to Intervene in a System (In Increasing Order of Effectiveness)

Note: Many of the terms to follow have everyday meanings, but also very specific but similar meanings in the world of systems thinking. If you want to know more, we very highly recommend you read the short paper *Places to Intervene in a System* which is widely available online.

12. Constants, parameters, numbers (such as subsidies, taxes, standards).

11. The sizes of buffers and other stabilizing stocks, relative to their flows.

10. The structure of material stocks and flows (such as transport networks, population age structures).

9. The lengths of delays, relative to the rate of system change.

8. The strength of negative feedback loops, relative to the impacts they are trying to correct against.

7. The gain around driving positive feedback loops.

6. The structure of information flows (who does and does not have access to information).

5. The rules of the system (such as incentives, punishments, constraints).

4. The power to add, change, evolve, or self-organize system structure.

3. The goals of the system.

2. The mindset or paradigm out of which the system — its goals, structure, rules, delays, parameters — arises.

1. The power to transcend paradigms.

No list is perfect or unmessy, as Dr. Meadows herself pointed out, and system drivers that appear to belong on one level of this list will sometimes ascend or descend to another level. Witness that California began requiring progressively higher energy efficiency with standards for appliances and buildings decades ago and to much howling resistance from industry, but today can boast a stronger economy and better homes and refrigerators, with lower per capita energy use — and the effect has rippled outward beyond California. What started as a simple set of rules became a wonderful positive feedback driver with no losers.

It very much bears pointing out that, as my old buddy David Eisenberg wrote after looking at a draft of this chapter,

> "We are actually dealing with multiple complex and interconnected systems, rather than a single complex system. I think it's a matter of which system, or which intersection of systems (for example natural systems like the climate system, or the financial economy, or the so-called regulatory system or systems) that we're considering that shifts the particular drivers up or down the list of effectiveness."

Or, as my Daddy always said, "*It all depends on how you look at it.*"

Depending on the context, codes and standards can be quite trivial, or a huge lever — and so can a price on carbon.

With that said, and with a loose eye toward Dr. Meadows' brilliant list, here are some suggestions in more or less ascending order of effectiveness for action — by individuals, by policy makers, by industry — for transitioning to a climate-friendly palette of building materials. Some would argue with the ordering or completeness of this list, and I would eagerly welcome that conversation, but few (I suspect) would argue with the assertion that carbon taxes and an informed populace (especially women and girls — see the end of this section) will be by far the strongest and most effective change agents not just for building but for climate action writ large.

Building Codes and Standards

In the century or so since its inception in human economy,[2] building regulation has been mostly geared toward keeping bad stuff from happening — injury, illness, death, and loss of property due to fire, earthquake, bad sanitation, and so on. A good and obvious first start! But now, here we are with the somewhat sclerotic codes ratcheting ever lengthier and more exacting, and making building ever more complex and expensive without necessarily adding much if anything to safety. We've got a good handle on fire, earthquake, ventilation, and sanitation, but how about all the new risks, like the effects our industry has on surrounding soil, air, water, and the air inside of buildings? How about the effects of our buildings on human and climate health? We're really pretty good at reducing the risk from all the obvious "first generation" risks, but so far turn a blind eye to a new generation of risk. It's time to address these concerns, but also write rules that encourage good things to happen (more on that in a moment).

We also want to emphasize that, though we place codes and standards here at the start, and thus by implication with the least leverage for change, *they are often much, much more.* Building and planning regulations can be massively influential for either the good or the bad. Witness US postwar urban planning that has baked in suburban sprawl and racial segregation, much to everyone's detriment, or the insidiously written laws requiring flame retardants in fabric and furniture that do little to curtail fire but produce serious toxic threats to many. On the positive side, a rule to constrain new construction's carbon footprint would prove to be a strong positive feedback driver. The power of regulations lies not so much in their letter as in the mindset that writes them, and the system of governance that publishes and enforces them.

Incentives

What if, for example, you could forgo permitting fees or property taxes as a reward for building below a set threshold of carbon emissions per unit area, or for using nontoxic, low-carbon or better carbon sequestering materials? What if you were rewarded for increasing density with second living units in suburbs or energy upgrades to old buildings, reusing existing urban sites rather than building anew at the urban perimeter?

Research

The USA lags far behind the rest of the industrialized world in fostering and sponsoring research that could make for better buildings of any sort, much less ones with less or better environmental impact. Whether from government, philanthropy, and/or industry associations, funding for science can yield enormous public and private benefit by highlighting bad practices (e.g., when higher insulation levels lead to moisture problems), or fuel innovation of new products (e.g., when better understanding of borax and borates made it possible to make fire- and bug-safe insulation from recycled newspaper). Just a few examples of research questions we'd like to see addressed would be:

1. How do we make plastics without all the horrible side effects discussed in Chapter Seven?
2. How can we make carbon-sequestering concrete?
3. Can we develop a family of inexpensive nontoxic binders by which to make artificial lumber products from straw and other agricultural residues?
4. How do we effectively harvest all the resources that are now diffused in landfills around the world?
5. How do we get the plastic out of the ocean and safely, effectively into building products?

Information Flows

Climate change is well documented and well studied (though there is of course a great deal more we'd like to know), and there isn't any rational doubt that it's happening or that it's because of us humans. However, climate disinformation is widespread and deliberate, especially in the USA where it has profound and unfortunate consequences; misinformed people elect misinformed leaders who make misinformed and far-reaching laws. Far more unfortunate is the emotional baggage that accompanies this disinformation, and rational conversation is increasingly rare and difficult in conferences, meeting halls, or family dinner tables. We can't make good decisions or build good buildings if we're not all working on the same reasonably complete information. (Far-fetched, yes, we know. As soon as we can find that magic wand that was here a moment ago, we'll wave it and turn everybody everywhere into educated, rational people eager to understand their opponents' point of view and move every conversation to a win-win conclusion. Heck, while we're at it, we'll make everybody deliriously happy, too. As soon as we can find that wand.)

A Price on Carbon

This is a big one, not necessarily because it looms larger than information flows, but because it's well within the realm of possibility. It is in fact already dawning in many places.

This entire book describes and promotes the transformation of buildings from climate *villains* — major CO_2e emitters — into climate *champions* — major CO_2e absorbers. There are already drivers in place to make this happen, as when mass timber structures are, sometimes, cheaper and faster to build than their alternatives of concrete and steel, or when a region can supply the insulation for buildings by skillfully utilizing farm and other by-products rather than imported oil.

But a hard truth still overarches this effort, and indeed all efforts to slow and mitigate climate disruption: cost. Almost everywhere in the world today, being a climate "villain" carries little or no cost, and being

a "hero" carries little or no reward. Money talks, and carbon keeps going up into the air. Carbon intensity is baked into the rules and rewards at almost every level of almost every part of our modern economy; we've got the fossil fuel habit, and we've got it bad.

Recognizing this, economists and policymakers are rapidly moving toward monetizing carbon in our industrial economy; to reverse climate change we have to make polluting hurt. This generally takes the form of carbon *taxes*, and/or carbon trading *markets* by which net emitters can buy credits from net absorbers so as to continue in business. For example, the cement plant, a major carbon emitter, buys credits on the market from the sustainable forester, a major absorber. The "sins" of one cost money, compelling him to move to cleaner operations, while the "virtues" of the other make her money, helping to clear a profit and expand her operation.

That's the theory, anyway. The reality, you will be shocked to hear, is substantially more complicated and, we almost need not mention, politicized. For example, how do you tax carbon without unduly burdening the poor, or punishing a business that has a huge capital commitment to "bad" technology? How do you price carbon so that a trading market can be effective? Even as carbon markets are rapidly proliferating across the globe, there are still many cases of ineffectiveness or outright failure. At present, trends favor carbon taxation over tricky trading markets.

The vast majority of governments around the globe — 189 countries representing

96 percent of global greenhouse gas (GHG) emissions and 98 percent of the world's population — have committed to reduce their GHG emissions and adapt to the changing climate . . . The urgent priority now is for governments to ensure implementation of these commitments, requiring sustained efforts to influence investment and consumption decisions made every day by firms and households.

The range of carbon prices across existing initiatives continues to be broad. This year, observed carbon prices span from less than US\$1/tonCO2e to US\$131/tonCO2e, with about three quarters of the covered emissions priced below US\$10/tonCO2e.

The World Bank report tells us that carbon markets are expanding, yet are still so immature that prices range wildly, and the markets are not yet having appreciable effect on "consumption decisions made every day by firms and households." Still, the trend is clear, and spawns a related trend of forward-thinking investors demanding that firms inventory their carbon footprints (itself an emerging science as has been described) and start preparing and planning for liabilities that may accrue as legislation at every level comes into alignment with the threat of climate disruption:

In 2016, the number of companies that are using an internal price on carbon has more than tripled compared to 2014. The internal carbon prices in use are diverse, with reported values ranging

from US\$0.3/tonCO2e to US\$893/ton CO2e. About 80 percent of the reported internal carbon prices range between US\$5/tonCO2e and US\$50/tonCO2e.[3]

The United States is second only to China as a global CO2e emitter, but has not been a leader, to put it mildly, in working to reverse climate disruption. The 2016 election will surely lead to even more intransigence at the federal level but, it is hoped, spur ever more determined and widespread efforts at other levels of governance in and beyond the USA. That, and the rising pressure on business to "clean up its act" or at least prepare for the costs that will come due for emitting, are the trends we will be watching. Because when it costs a lot to emit — when somehow the emitters can feel the hurt — and conversely makes money to soak up carbon, then all of the technologies described in these pages, and many more yet unimagined, will receive an enormous boost to competitiveness and viability.

Necessary Afternote #1

We offer this frighteningly short summary of a complex subject — carbon markets and pricing — because the book calls for at least a cursory discussion. Needless to say, it gets more complicated and more controversial than we've had room to discuss, because it is a fast-changing and evolving issue with a lot of money involved. When there's lots of money involved, people's eyes get narrow and they reach for their guns, metaphorically if not literally — not an easy setting for rational discussion. We would further add as:

Necessary Afternote #2

In response to the snorting and harrumphing we can hear from our colleagues of libertarian ilk, bellowing about a free market: bite us. We too would be curious to see a free and unencumbered market at work, but to our knowledge, it has never been tried, not even close, and in our expectation will likely *never* be tried. Humans are humans; we'll game or cheat or rig the system if we possibly can, party animals that we are. More to the point: the economy that we now have in the industrialized world is based, for better and worse, on cheap and abundant fossil fuels but *not* a free market. Rockefeller provided the gas, but the taxpayers provided the roads, and today pay for the US Navy in the Strait of Hormuz. The US Department of the Treasury estimates that direct subsidies to the fossil fuel industry total about $4.3 billion per year in the US[4] while the International Monetary Fund estimates that total direct and indirect subsidies to the fossil fuel industry, worldwide, total about $5.3 *trillion* USD per year.[5] The fossil fuel economy has enjoyed its two hundred years with a stacked deck, now let's transition to an economy favoring clean energy, air, soil, and water.

Necessary Afternote #3: Which System Are We Talking About?

To build (or to do most anything else of any size) is to work within systems of systems of systems — materials within products within assemblies within buildings within labor and commercial markets within local and national governance within a world economy. To a very large extent, all of these levels understand themselves in purely economic terms; they don't recognize that each and all are very much subsidiary to the overarching Big System: Life on Earth. For centuries we have externalized Nature as some vast thing apart, an "over there" from which we could extract stuff we wanted and dump stuff we no longer wanted. We got away with that when there weren't so many of us and we didn't push the enormous resilience of Life too far. Now we are a still-expanding species of seven billion and pushing past many limits. Putting a price on our carbon emissions is a clear and effective way of establishing a more viable relationship with the Big System on which we depend.

Necessary Afternote #4: In Which the Republicans Make the Case

We close this argument with an extended quote from the Climate Leadership Council,[6] an august collection of US businessmen and former Republican federal officials including such wild-eyed liberals as James Baker III, Martin Feldstein, Henry Paulson Jr., George Shultz, Thomas Stephenson, and Rob Walton:

> *Mounting evidence of climate change is growing too strong to ignore. While the extent to which climate change is due to man-made causes can be questioned, the risks associated with future warming are too big and should be hedged. At least we*

need an insurance policy . . . Economists are nearly unanimous in their belief that a carbon tax is the most efficient and effective way to reduce carbon emissions. A sensible carbon tax might begin at $40 a ton and increase steadily over time, sending a powerful signal to businesses and consumers, while generating revenue to reward Americans for decreasing their collective carbon footprint . . . trade remedies could also be used to encourage our trading partners to adopt comparable carbon pricing . . . Our reliance on fossil fuels contributes to a less stable world, empowers rogue petrostates and makes us vulnerable to a volatile world oil market. Carbon dividends would accelerate the transition to a low-carbon global economy and domestic energy independence. Not only would this help prevent the destabilizing consequences of climate change, it would also reduce the need to protect or seek to influence politically vulnerable oil-producing regions. With our electric grids susceptible to cyber attacks, a transition to cleaner power sources combined with new distributed storage technologies could also strengthen national security.

But the biggest driver of all, as it turns out, is women.

Long after being birthed by one, but well before I married one, raised one, and in general have been blessed with all manner of rich relationships with women, I had begun to suspect that they are key to the transition and healing we're talking about. That notion was solidly reinforced just as we were finishing the draft of this book. My wife Sarah and I attended a talk by my old colleague Paul Hawken introducing his exciting new book *Drawdown,*[7] which summarizes an enormous worldwide research effort to identify the most promising ways to reverse climate change. Their conclusions are inspiring and full of surprises (read it!), but also support the premise of this book: we can greatly reduce emissions and sequester a huge amount of carbon in our built environment — and also get much better buildings for the effort.

In Paul's list of the top ten drivers to draw down atmospheric carbon (the full list is a hundred items long), a few surprises occupy the top spots. But, if you combine two closely related items just down the list, educating girls and family planning, you get an effect that dwarfs any single other item. This may not have much to do directly with architecture and building materials, though women do show up much more in natural and green building than the industry as a whole. But I'm pretty sure it has everything to do with a cooler climate and a cooler world, and we couldn't close this book without a tip of the hat and one deep bow. It may look like it's about the solar panels and the concrete, but it's not, not in the long run. It's about those who have a much better track record than some of these other genders in nurturing children, building community, and husbanding resources. They may not be as good, generally, at moving furniture or hot dog eating contests, but it's looking more and more like it will be women who guide us home.

Notes

1. http://donellameadows.org/archives/leverage-points-places-to-intervene-in-a-system/
2. (with a respectful nod to Hammurabi, the Babylonian king who instituted the first, though somewhat harsh, building code in history almost four thousand years ago. "Build it right or lose your life' seems like a fairly strong system driver!")
3. Quotations from *State and Trends of Carbon Pricing* by the World Bank Group, October 2016.
4. http://g20.org.tr/wp-content/uploads/2015/11/Summary-of-Progress-Reports-on-the-Commitment-to-Rationalize-and-Phase-Out-IFFS.pdf
5. www.imf.org/external/pubs/ft/survey/so/2015/NEW070215A.htm
6. www.clcouncil.org/wp-content/uploads/2017/02/TheConservativeCasefor CarbonDividends.pdf
7. www.drawdown.org

Afterword

If we want things to stay as they are, things will have to change.

— *"The Leopard,"* Giuseppe di Lampedusa

W<small>E HAVE PRESENTED</small> this as a book about architecture and building materials, but it is also an invitation. Not to take anything from myself or my co-authors, for we have covered a great many technical topics, briefly but, we hope, with clarity and competence. To make for a simple read, we chose to barely touch on a great many subjects (and surely omitted a few that deserved attention) that each call for more notice, if not an entire book, to themselves. But this is essentially an excited message, a wild wave of the arms, from us to you, good news announcing a major part of solving a problem such as human beings have never faced before. So far as anyone can discern, there has never been a moment in history anything like this. It is terrifying, it is exhilarating, and it is one hell of a time to be alive. It is in that spirit that we present the new carbon architecture, a first sketch of an emerging picture: we can now cool the climate and live in buildings made, literally, of sky.

In the preface, I told of seeing an electric car with the license plate "ZERO CARB,"

presumably boasting of its light carbon footprint — a boast that I derided but also declared to be the catalyst for this book. Lo and behold, some months later as we were nearing completion of the draft, I passed another Tesla with the license plate "FRE NRG," articulating in six letters the bookend myth of the green movement, and really of our entire culture.

Call me a party pooper, but there's no "zero emissions" and there's no "free energy." Everything we do has effects, some of which we see, some of which we don't. And everything we do takes energy. Even if it's the abundant free energy of the sun or wind, we expend energy to harvest it. That's the way things work in this world, but that's not something to bemoan or evade — it's something to celebrate and work with! We live and build in a world of endless change and exchange, energy and matter shifting form at all sorts of time and size scales — from microseconds to millennia, from cellular to global. We're no more capable of understanding it than a frog is capable of chess — yet that is exactly what we must try and do. Earlier we quoted Janine Benyus describing natural systems as "resilient, decentralized, self-repairing and scalable without experiencing complexity problems . . . and [able to] plug leaks and make use of materials previously considered to be waste." We submit that architecture and construction need to start emulating those proven concepts such as: no waste, no hidden effects or sneaky unaccounted use of natural capital, and plenty of exuberance. Some call this a circular economy, but it's more like a cosmic dance, circling among circles within circles around circles, all swirling amongst, over and under themselves according to rules we're only beginning to understand. Nature's way is just more fun.

In India there is a well-known story about Shiva, the God of Destruction (though he has many other dimensions, as this story illustrates). It seems that once upon a time a bunch of gods got together and, peeved with humanity, decided to dump a particularly nasty poison in the ocean that would wipe out all of life. Shiva got wind of this plot, dove into the ocean, and swallowed all the poison so as to save the human beings, which is why he became blue and also why he is so revered by so many. My old teacher P.K. Mehta told me this story years ago, and

opined that the construction industry could be metaphorically like Shiva, absorbing much of the toxic junk we've been producing over the past century and turning it into fine architecture. Sort of an ecological triage until we stop producing trash and poisons, and get the hang of working with, not against, Nature. We've been messing with the climate, and now must clean up that mess. It's just civilized behavior, after all, to clean up after yourself before you cause problems for others. Also, not to belabor a point, being truly civilized is more fun.

And speaking of fun, consider carbon, the "Let's party!" molecule. Give it some hydrogen, oxygen, nitrogen, a bit of iron and silicon, and next thing you know you've got fields of daisies, pods of whales, the Amazon rainforest, and you, reader, with your carbon eyes and carbon brain taking this in. The trick before us is to get more of it out of the air and back on the ground as schools, offices, and homes — a new carbon architecture. And so here you have it, a book made of carbon, written by carbon beings for other carbon beings, on how to entice carbon from the air to build carbon shelter to protect us from a sometimes hostile carbon planet. It's carbon all the way, baby, a carbon symphony with a funky beat gets you on your feet.

Shall we dance?

Contributing Authors

In order of appearance:

Bruce King is the founder of the Ecological Building Network (EBNet), and a registered engineer with 35 years of worldwide experience in structural engineering and construction. He is the author of *Buildings of Earth and Straw* (1996), *Making Better Concrete* (2005), *Design of Straw Bale Buildings* (2006), ASTM International E-2392, earthen building guidelines, and dozens of papers and articles for conferences and journals. He has organized three international conferences on ecological building, and is the founder of BuildWell Source, a user-based collection of low-carbon materials knowledge, and of the BuildWell Symposia. www.bruce-king.com www.ecobuildnetwork.org

Erin McDade is a Program Manager for Architecture 2030. She holds a Master's of Architecture degree and previously worked at the Integrated Design Lab. While with the IDL, she helped develop Targeting 100!, a tool enabling deep energy retrofits and high-performance new construction in the healthcare sector. She also completed lighting and thermal analyses of the revolutionary Bullitt Center. She leads Architecture 2030's Products Challenge, is founding chair of the Embodied Carbon Network, and is on the board of the Carbon Leadership Forum. She is also leading AIA+2030 Online Series development, helping design professionals create zero carbon buildings.

Ann Edminster is a leading international expert on zero-energy efficient green homes. Founder and principal of Design AVEnues LLC, Ann consults with builders, developers, homeowners, supply chain clients, design firms, investors, utilities, public agencies, and non-profits — from local to international organizations. Her award-winning 2009 book, *Energy Free: Homes for a Small Planet,* is a comprehensive guide for designers and builders seeking to create zero net energy (ZNE) homes. She assists design teams in pursuing ZNE performance goals, has developed curricula for design and construction of ZNE homes, and is a frequent keynote speaker, presenter, and teacher at conferences, universities, non-profits, and utilities.

Catherine De Wolf is a postdoctoral scientist working on low carbon structural design at the Swiss Federal Institute of Technology in Lausanne. She obtained her PhD at the Massachusetts Institute of Technology (MIT) and worked at the University of Cambridge in the field of embodied carbon. She holds a Master of Science in Building Technology from MIT and a Bachelor and Master of Science in Civil Engineering and Architecture from the Vrije Universiteit Brussel and the Université Libre de Bruxelles, Belgium. She has spoken about low carbon building materials at TEDx in Paris and when receiving the Innovators Under 35 Award in Belgium.

Kathrina Simonen is an Associate Professor in the Department of Architecture at the University of Washington with over 15 years of professional practice experience as an architect and structural engineer. Her research is focused on understanding and reducing the environmental impacts of manufacturing building materials and products through the use of environmental life cycle assessment (LCA). She is founding director of the Carbon Leadership Forum, an industry-academic collaboration focused on linking LCA to design and construction practice and has authored a handbook, *Life Cycle Assessment,* a primer for building industry professionals looking to learn about LCA.

Barbara Rodriguez Droguett has devoted over a decade to the creation and improvement of analytic tools to assess carbon in buildings. She was the first LEED AP woman in Chile and later the Chief Sustainability Officer at the Center for Innovation and Research of Buildings at Universidad de Chile, where she led the first EPD program for the building sector in Latin America. She was director of ECOBASE, the first nationwide LCI for building materials in Chile. In 2015 she received the National Award for Sustainable Construction Leaders under 35 from the Chilean Chamber of Construction. She is currently pursuing a PhD at the University of Washington.

Larry Strain, Siegel & Strain Architects. Larry has a 40+ year background in sustainable design and studied ecological systems at Evergreen State College and the Farallones Institute. He wrote a Guideline Specification for Green Materials, which became part of Building Green's *GreenSpec Directory*. He has spoken on materials and sustainability at conferences throughout the country, is a past board member of the Northern California Chapter of the USGBC, and currently serves on the boards of the Ecological Building Network and the Carbon Leadership Forum. For the past seven years, Larry has focused on reducing the total carbon footprint of our buildings.

Frances Yang is a structures and materials sustainability specialist in the Energy + Sustainability group of the San Francisco office of Arup. Frances uses her studies in structural engineering, life cycle assessment, architecture for the environment, and green chemistry in leading the Sustainable Materials Consulting practice for the Arup Americas region. She recently served as vice-chair of the Materials and Resources TAG of USGBC and chair of the ASCE/SEI Life Cycle Assessment working group. She also contributes to the Carbon Leadership Forum and AIA Materials Knowledge Working Group. More info on Arup Materials Consulting at www.arup.com/services/materials

Andrew Lawrence is the leading timber specialist at Arup, a member of the European Timber Design Code Committee, and a judge for the UK Wood Awards. Andrew lectures worldwide on the structural use of timber and is currently working with timber industry bodies in the USA, UK, China, southeast Asia, and Australia, to help make timber a mainstream construction material. His projects include the highly acclaimed Metz Pompidou, the Canary Wharf Crossrail Station roof, and most recently The Smile, the world's first hardwood CLT structure.

Jason Grant has been a leader in the sustainable forestry and green building movements for 25 years. Jason co-founded EcoTimber, one of the first companies in the world to bring certified ecological forest products to market. A LEED AP Building Design and Construction, Jason is recognized as an expert in ecological forest products and their role in green building. He has long advocated for sustainable forestry and responsible wood use as a member Sierra Club's Forest Certification and Green Building Team. He is the author of *(R)evolution in the Redwoods*, an educational program on the transition of California's redwood industry from exploitation toward restoration and stewardship (www.redwoodevolution.com).

Chris Magwood is obsessed with making the best, most energy-efficient, carbon sequestering, beautiful and inspiring buildings without wrecking the whole darn planet in the attempt. Chris is currently the executive director of The Endeavour Centre, a not-for-profit sustainable building school in Peterborough, Ontario. He has authored numerous books on sustainable building, including *Essential Hempcrete Construction* (2016), *Making Better Buildings* (2014), and *More Straw Bale Building* (2005). In 1998 he co-founded Camel's Back Construction, and for over eight years helped to design and/or build more than 30 homes and commercial buildings, mostly with straw bales and often with renewable energy systems.

Massey Burke is a natural materials specialist in the San Francisco Bay Area. Her work centers on research, design, and hands-on implementation of building with low-carbon natural materials, with an interest in applying natural building to existing buildings and the urban fabric. She also teams up with organizations such as the California Straw Building Association and the Ecological Building Network to generate technical information on carbon and natural materials in the built environment, and works on code issues surrounding natural materials through the Cob Research Institute and other collaborators. More on her work can be found here: http://masseyburke.carbonmade.com.

Craig White is a developer, architect, and entrepreneur. Founding Director of White Design with over 25 years' experience in architectural practice in the UK and Europe, Craig is also the Director of ModCell Straw Technology and Coobio Circular Materials. Craig is currently leading work on a new model of community-led and -financed housing that meets the housing crisis challenge using carbon-banking renewable materials. Craig is also a consultant with the Carbon Trust, and senior lecturer at the School of Architecture and Planning at the University of West England. A core focus of Craig's work is how research-led innovation delivers commercial impact and sustainable outcomes.

Fernando Martirena is the director of CIDEM (Center for Research & Development of Structures and Materials) at the Universidad Central de las Villas in Santa Clara, Cuba, which is a world-leading institution in the development and implementation of appropriate technologies for social housing. Professor Martirena spearheaded the emergence of appropriate technologies in Cuba and Latin America at the beginning of the 1990s, with a strong focus on making cement more sustainable through the use of pozzolans. Through the years, this movement has evolved to the award-winning Latin American Network for the Sustainable Habitat, Ecosur (ecosur.org), that encourages south-to-south technology transfer.

Paul Jaquin is a chartered structural and geotechnical engineer working in New Zealand. Paul completed his PhD thesis, entitled "Analysis of Historic Rammed Earth Construction," in 2008 at the University of Durham, UK. Working as a consulting engineer, he has undertaken the design of a number of earth buildings. Paul is a highly cited author of academic papers and books regarding earth building, and has spoken at conferences around the world. www.historicrammedearth.co.uk

Mikhail Davis is Director of Restorative Enterprise at Interface, the world's largest manufacturer of modular carpet. He is responsible for advancing Interface's globally recognized Mission Zero commitment in the Americas by building internal leadership capacity and creating external partnerships. He also chairs the LEED Materials and Resources Technical Advisory Group for the US Green Building Council. Previously, he served as manager to environmental icon David Brower and spent five years with Blu Skye Sustainability Consulting building sustainable business strategies for Fortune 500 companies. He holds a BS in Earth Systems from Stanford University and is a certified Biomimicry Specialist. www.linkedin.com/in/mikhail-davis-661bb4

Wes Sullens is an advocate for circular material economies and a regenerative built environment. He has worked in the public, private, and non-profit sectors on a range of topics including waste management, recycling, supply chain sustainability, and chemicals transparency. He specializes in green building rating system development, product labeling standards setting, and progressive green building codes advocacy. Wes is a LEED Fellow and currently works for the US Green Building Council as their Director for Building Codes Technical Development.

Wil V. Srubar III, is an assistant professor of architectural engineering at the University of Colorado Boulder (CU). A structural engineer by training, he received his PhD in 2013 in Civil and Environmental Engineering from Stanford University with a specific focus on biopolymers and bioplastics for construction. At CU, he is actively engaged in research projects related to durable, low-carbon polymer- and cement-based construction materials. He is an active member of the American Society of Civil Engineers and the Architectural Engineering Institute, and he currently serves as a co-chair of the Embodied Carbon Network. spot.colorado.edu/~wisr7047/

Pete Walker is Professor of Innovative Construction Materials and Director of the BRE Centre of Innovative Construction Materials at the University of Bath, UK. A Chartered Civil and Structural Engineer, Pete has been doing research on construction materials and technologies for over 30 years. His particular interests are natural materials, including straw bale, earth building and natural fiber composites. He has published over 180 articles. He teaches in undergraduate and postgraduate programs in the joint Department of Architecture and Civil Engineering at the University of Bath. He was the inaugural chair of Earth Building, UK, the national association for earth building.

Andrew Thomson works at the University of Bath as a Research Associate. His work focuses on advancing the use of low-carbon construction materials within the construction industry. He has contributed to the structural design of some of the UK's largest Cross Laminated Timber (CLT) buildings and was a key member of the research team that delivered the UK's first certified straw bale panel product; ModCell Core. Andy has also worked as a rammed earth building contractor in New Mexico and as an Associate Lecturer at Oxford Brookes University.   www.bath.ac.uk/ace/people/thomson/index.html

Daniel Maskell is a Prize Fellow in Innovative Construction Materials in the Department of Architecture and Civil Engineering at the University of Bath, UK. Dan's interests are in innovative building materials and how these can be used for the improvement of indoor environment quality for improved occupant health and well-being. He has wide experience with natural building materials, including earth, straw bale, and other inorganic and organic materials. Dan's approach is to consider holistically the use of these materials, investigating their material properties for structural and indoor environment regulation.

Index

A Note about the Publisher

NEW SOCIETY PUBLISHERS is an activist, solutions-oriented publisher focused on publishing books for a world of change. Our books offer tips, tools, and insights from leading experts in sustainable building, homesteading, climate change, environment, conscientious commerce, renewable energy, and more — positive solutions for troubled times.

We're proud to hold to the highest environmental and social standards of any publisher in North America. This is why some of our books might cost a little more. We think it's worth it!

+ We print all our books in North America, never overseas
+ All our books are printed on **100% post-consumer recycled paper**, processed chlorine free, with low-VOC vegetable-based inks (since 2002)
+ Our corporate structure is an innovative employee shareholder agreement, so we're one-third employee-owned (since 2015)
+ We're carbon-neutral (since 2006)
+ We're certified as a B Corporation (since 2016)

At New Society Publishers, we care deeply about *what* we publish — but also about *how* we do business.

New Society Publishers
ENVIRONMENTAL BENEFITS STATEMENT

For every 5,000 books printed, New Society saves the following resources:[1]

28	Trees
2,502	Pounds of Solid Waste
2,753	Gallons of Water
3,590	Kilowatt Hours of Electricity
4,548	Pounds of Greenhouse Gases
20	Pounds of HAPs, VOCs, and AOX Combined
7	Cubic Yards of Landfill Space

[1]Environmental benefits are calculated based on research done by the Environmental Defense Fund and other members of the Paper Task Force who study the environmental impacts of the paper industry.